The Other End of the Stethoscope
(Tales and Lessons from Cancer, Women's Health Care, and Medical Education)

The Other End of the Stethoscope
(Tales and Lessons from Cancer, Women's Health Care, and Medical Education)
By Larry E. Puls M.D.

Foreword and Editorial Assistance
By Joel Oliphint

Foreword

The Puls family moved away from Amarillo, Texas, before my family did, but we soon followed suit—the Oliphints in Philadelphia, and the Puls eventually settling in Greenville, South Carolina. But that didn't stop us Oliphints from taking road trips down I-95 and I-85 to descend upon them.

I looked forward to those childhood vacations to Greenville. I remember each time we went, Dr. Puls had taken on a new hobby or project. At one point it was ceramics, making bowls and such out of clay. Another time we arrived to find a vineyard taking up a good portion of the backyard. (He dubbed the resulting wine "Chateau Lars.")

And, of course, there were the competitive sports. Larry had created quite an impressive disc-golf course on the property. We'd play nightly if the weather cooperated, and Larry would win every time. The man can toss a Frisbee with stunning precision. He was also undefeated on the ping-pong table, even while playing with a book instead of a paddle to make it more fair.

It was next to that ping-pong table in an upstairs rec room that my brother and I had our first conversation with Dr. Puls about what he did for a living as a gynecologic oncologist, which led, naturally, to a discussion of cancer. I knew what cancer was, but I never knew exactly how it worked, what damage it caused, how it spread, etc.

I remember walking away from the conversation with a much better understanding of cancer, but even more so, I was struck by how this man was doing exactly what he should be doing in life. Larry's fascination with cancer just oozed out of him. He couldn't contain it. He'd be downright giddy talking about massive tumors.

And yet, at the same time, he made sure we knew cancer was serious business. It can wreak just as much havoc on marriages and families as it does on a person's body. Some patients are spared, but many are not. Listening to Larry speak about it, there was no doubt that he took his responsibility seriously and that he consistently gave his all for every patient.

You'll see all of that in this book, a memoir that goes a long way in humanizing medical professionals. You'll watch as he tries not to get too emotionally attached to terminally ill patients, and then slowly relents and lets them into his life as they rope him into theirs. You'll squirm as he approaches a patient and her family to tell her that, no matter what, this cancer will eventually take her life. And you'll breathe a sigh of relief when a treatment regimen sends a cancer into remission. There's giddiness, and there's gravity.

He doesn't shy away from the quirky questions either, such as, "Do doctors have favorite patients?" (Yes.) Or, "What kind of music do you listen to in the operating room?" (See the "Beethoven" chapter.) His years of experience have also put him in quite a few outrageous situations, like keeping watch over a hallucinating man in a straightjacket, finding a pair of dentures on an incision and dodging bullets. (Not the metaphorical kind. Real bullets.)

As a grown man with a family of my own now, Larry's experiences with cancer and the patients he cares for still amaze me. And while I'm pretty sure he could still pursue alternate careers as a ceramist, vintner, disc-golf pro or table-tennis Olympian, this book will make you glad he chose the path he did.

Dedicated to Mary Ellen

This work is dedicated to my wife and forever companion. For all the long nights of studying, falling asleep in my lasagna, countless missed meals, multiple moves all over the country, and all the things I failed to get done at home while dealing with work, there could be no words of thanks that could adequately express my gratitude and love.

Table of Contents

A Very Special Patient 1
Size Matters 11
In Harm's Way 21
Almost Cut Down by a Logger 29
Beethoven 37
A Tale of Two Husbands 45
A Resident Distress 53
She's Greek to Me 63
Hallowed Halls 73
Bad News Bearer 83
Southern Belle 91
Back to life 99
Taken Down on the Elevator 109
Spoiled Brat 115
The Unexpected in the OR 125
The Pursuit of Knowledge 137

Chapter 1

A Very Special Patient

Have you ever wondered if your doctor has someone at the top of the "very special patient" list? A patient who consistently affected your physician in a positive way and inspired him, or just made a great day out of a dull day? And if that patient exists, did you ever wonder about the reasons this person was so liked?

In the twenty-something years I have practiced medicine, I have had the opportunity to care for a lot of wonderful people. They have come from all walks of life. They've been young and old, fluffy and thin, funny and serious. So as I recount the story of a most special patient, it is only to relate the endearing factors she had that I admired.

When I met "Jennifer," she was about 50 years old and in normal health when she noticed a new bump in her groin area. She didn't think much of the bump at first, but then it became annoying and wouldn't go away, so she eventually went to see her doctor. A CT scan revealed yet another mass that was a pelvic tumor, and unfortunately it looked like an ovarian cancer. Her surgery was scheduled and performed by my partner. Once her abdomen was opened and the mass removed, the pathologist called back with an ominous report. The tumor was malignant, and it had spread to a lymph node (the initial bump). Jennifer had ovarian cancer.

A few days later I took over on-call duties and met her in the hospital. She was recovering well. It was in this immediate post-operative period that I talked with her for the first time. I was struck by her sense of humor balanced with the reality of the cancer. She recovered from the surgery and was released from the hospital.

Once the diagnosis of ovarian cancer is given, most women, with few exceptions, will end up on chemotherapy. The chemotherapy is usually given every three to four weeks and is administered over about a six- to seven-hour time span. This is done over a period of about six months. During that time, it's easy for an oncologist and the chemo nurses to form a bond with the patient and her family. It's also not unusual for relationships to form between chemo patients, since many of them are on similar treatment schedules. So while the patient is in the chemotherapy suite receiving her treatment with her friends, there's a lot of gossip.

Gossip, you say? Well, if there's one thing I have learned in life, it's that women tend to enjoy conversation. I think I can say that safely since all my patients are female, all my office staff is female, and I have a daughter and a wife. They have all taught me much. And they all love to talk and discuss the topic du jour. These patients are no different. Once they're comfortable with each other, they discuss life, chemotherapy side effects, war stories and the like. These discussions carry over to chemotherapy nurses. The chemo day becomes one big gab session. This gab session then becomes a gossip session. When I walk through the chemotherapy suite, I am always at a definite disadvantage since these ladies have already discussed everything under the sun, including things about my own life. Maybe my clothes don't match. Maybe, for once, they do match. Who knows? It seems all conversation is fair game.

I understand this concept firsthand. My wife is a breast cancer patient and had to receive radiation after her surgery. She would go down to the radiation area at the same time every day. Most of the other patients there were men getting radiation for prostate cancer. As they congregated at the same time every day, they began to know each other well. First they learned about each others' families, and eventually they learned about each others' idiosyncrasies. My wife even had a shirt made that said, "I'm radioactive." That was worth a lot of laughs with the guys. She was basically the only woman at that time slot, so can you imagine how much a bunch of older guys enjoyed conversing with a younger woman (and cute I might add).

After Jennifer had round one of chemotherapy, she decided it was time to spice things up a bit during treatment--give something for the girls to talk about. She was never satisfied to be just like everyone else. It wasn't enough just to get chemotherapy; it had to be dramatic. So she came in early that morning and decorated her little corner of the world. It was dressed up with banners, colored confetti, and completed with a "barf bucket." Now, she never actually threw up while getting chemotherapy (at least that I remember), but the barf bucket was certainly an interesting little touch. That little touch led to a lot of discussion and a lot of laughs. Soon enough, the barf bucket, the decorations and Jennifer herself were infamous.

I think these props helped maintain her sanity in the topsy-turvy world she found herself in. Not to be daunted by something so trivial as ovarian cancer, she sailed through her six initial rounds of treatment. The greatest blessing of all was that, at the end of the six months, she went into remission. Tears welled up in her eyes. She had taken an awful diagnosis and showed it who was boss. There was no doubt in my mind she wanted to rule this cancer, and she was off to a great start. During the ensuing months, she and I talked often and traded war stories on her visits. I could always tell, however, that hidden down inside that tough exterior were lots of worries and consternation over her examinations and blood tests. She knew that it could come back.

A little more than a year later she developed new signs of the cancer. The CT scan was positive, and the dreaded words came from my mouth: "The cancer has returned." Once ovarian cancer comes back like this, death is usually inevitable. There are few exceptions. And yet let me qualify that by saying that even though cures are rare at this point, oncologists now have so much to offer in this world of recurrence. There are surgeries, occasionally radiation, and many active chemotherapy drugs that show very real promise. I have more and more patients that survive many years after their initial recurrence, whereas 20 years ago, I almost never saw patients with recurrent disease survive more than two years. Sometimes we look at ovarian cancer as a chronic disease.

While going through the list of options in Jennifer's case, chemotherapy was the most logical and reasonable course of action. In Jennifer's previous six months of chemotherapy, I saw her showmanship. She dazzled with her sense of humor and over-the-top antics. She was one eternal bag of surprises. During her recurrence, I saw a new side to her, something I can only describe as her "lawyerness." As she started down the pathway of chemotherapy for a second time, she became a student of oncology, learning much about each and every chemo drug. In some ways she was more knowledgeable about the side effects of the drugs than myself because she was the one actually taking the chemotherapy. What I observed and memorized about side effects doesn't compare to actually experiencing those side effects in person. As I would teach and instruct her about each new drug, she was a sponge, memorizing and noting everything. Then she would experience the drug first hand. Since she was armed with information about each drug and each side effect, she began to correlate this with the dosing of the drug. Now the effects had meaning to Jennifer based upon the actual dose that she received.

So the lawyering began. She would look at her schedule each month and tell me her social calendar. Depending on what she was doing that month and on which day the event was occurring, she would debate with me the timing of the chemotherapy, the type of chemotherapy, and the dosing. For instance, she would say, "I have a party 12 days from now. If you give me drug X and it knocks down my white counts, I might get sick at my own party. Since my CA-125 [the blood test for ovarian cancer] counts are better today, could we drop the dose to Y and delay giving it by three days?" The first time she said something like that to me, I didn't know how to take it. No one, and I do mean no one, had ever asked me something like that. I was taken aback.

But that's just how she was. At times it would aggravate me. Other times I had to laugh. This became her regular habit. Her social calendar was full and she was not going to be controlled by her cancer. She wanted to have some control over it. She believed the cancer shouldn't drive every decision in her life. I can appreciate that. It was good for her mentally, I think, to view her life through that lens. After arguing against her for a while, I finally began to acquiesce. It sometimes just seemed easier to give in than to fight it. Obviously, if I thought the idea was not acceptable, I would not do it. But because she became so well versed in the art of delivering chemotherapy, it was sometimes difficult to debate with her. Still, she was always gracious about it, putting a great, big smile on her face and asking, "Please?" How do you say no to a woman who is dying from ovarian cancer, who asks intelligent questions, and then smiles at you while waiting for you to grant her request?

Jennifer also had an endearing ability to roll with the punches. If things didn't go her way, she would write it off as part of life. Nothing represented this quality so well as one particularly unusual event involving a Port-a-Cath, which is a device placed into patients to gain IV access. When people are on chemotherapy, they end up having so many IVs and blood draws that eventually they run out of viable veins. The blood sticks get harder and harder and the veins become more difficult to find. Great IV access is so critical to these patients, in fact, that we call their lines "lifelines." The name seems appropriate because without them, they can't get chemo, and without chemo, they cannot stay alive. Good access is also critical because if veins get weak and the chemotherapy leaks out under the skin, the area where it leaks will begin to die and large necrotic ulcers appear. So great IVs are critical to treatment.

Jennifer had such an access. We placed it in after she began to run out of veins. The access had been used for a number of months, and then suddenly it stopped working. The nurse called me and asked what we should do. We generally obtain some X-ray studies to determine the line's viability. It's a pretty routine test. On this day, however, I got the most bizarre phone call. The radiologist called to say that not only was the line not working, but the catheter had broken off. After bending her arm umpteen million times, the Port-a-Cath line severed in half and dislodged into her heart. In 20 years of doing this job, and about a million Port-a-Caths, I had never seen this happen. Never.

I asked the radiologist what we could do. He thought he could retrieve it right out of her heart. Unbelievable? I thought so, but by George, he did it. The interventional radiologist had to go in the right side of the heart, visually find this moving target, grab it, and pull it out. And that's what they did. Jennifer's reaction?

"If it is going to happen, it will happen to me."

She began to tell me how every weird thing that can happen on Earth always seemed to happen to her. In my short time of knowing her, I came to believe that. Who was I to question it? Her life did seem strange and unusual. When I saw her after they retrieved the catheter, she just shrugged her shoulders and asked where we went from there. It was as if nothing had happened. She didn't care. I got the sense that this was just another chapter in her strange life.

It would be unfair to recount this story without telling of her husband. She was indeed blessed with a wonderful man in her life. He came to at least 90 percent of her office visits. When she would negotiate with me about the timing and dosing of the chemo, he would just say, "Don't look at me. She's the demanding one." He would tell me that it was the same way at home. They had a life of mutual admiration and love. They took dates all the time and went away on trips together commonly. It was not unusual, in the midst of all her suffering and struggles, that she would ask me about her physical relationship with her husband. She would say they had a big weekend away planned and wanted to know if she could, well, be intimate. It would do my heart good to see this kind of support on both ends of the relationship. When they took the vows, they lived them out. I've seen marriages destroyed by cancer. I've seen bitterness develop, mostly over selfish reasons, on both sides. But I must say, it was a pure joy to see this kind of marriage lived out before my eyes. That is one of the blessings of my job. I get a window seat next to many lives. Sometimes I like what I see through that window, and sometimes I don't. This time, it was beauty in motion.

Eventually, she began to deteriorate. She took the same course so many ovarian cancer patients take--one of a high-grade blockage of the intestinal tract. The patients finally lose the ability to eat, which is an awful thing. Eating is such a blessing. We could be like the cattle and only eat grass. But the good Lord has given us so many great things to eat. Great foods. Great spices. Great beverages. Losing the ability to consume these blessings is a terribly difficult situation. But I watch this event in so many patients with ovarian cancer. Death by bowel obstruction is the number one cause of death in these women. It is as if a vice grip has been placed on the bowel and slowly the women lose the ability to consume solid food. Then liquid food. Then ultimately they can consume nothing at all. Starving to death is beyond my comprehension, yet I see it all the time. It never gets easier to witness.

Around this time in treatment, I usually begin to withdraw from face-to-face care with patients. It's a defense mechanism that I have purposefully constructed. You see, I can't die with all my patients. If I did, I would never allow myself to get close to people, because I know that most patients with stage III ovarian cancer will die. Not all of them, but many of them. Knowing this, I build into my life these certain safeguards that don't allow me to die with every patient. One thing I am very thankful for is the Hospice program. This program is designed to let me manage my dying patients from afar. Although you may not understand the reasoning, I cannot stay in the trenches face to face with all of my dying patients. If I did, I'm sure after a while I would cease to enjoy each patient situation. It's one thing to have a friend die, but it's an entirely different thing to have 50 friends die every year. Either I would get numb to life, or I would go insane. What I choose to do is stay involved in each precious life, but begin to withdraw at the end. In the world of oncology, you need defense mechanisms. It keeps my sanity intact, and it keeps me involved.

That said, I stayed involved in Jennifer's care face to face longer than most patients. I think it was partly because she had the fortitude to continue to come to the office even in the waning days of her life, and also partly because I had come to enjoy her friendship. The many days of negotiating doses of chemotherapy made me appreciate her resolve. The many days of walking by her decorated corner of the chemotherapy suite made me appreciate her sense of humor. The many days of talking to her about her social calendar and her dates with her husband made me appreciate the human experience. We taught each other many things. Our relationship, I believe, was one of mutual admiration.

Jennifer's final visit to the hospital came like a thief in the night. After many years of managing her ovarian cancer, I knew that death was slowly coming, I just didn't know when. She called me one day saying she couldn't eat. She was throwing up everything she tried to ingest. The nausea was unrelenting. So I admitted her. After all of my tests were back, the conclusion was obvious. She was blocked. This was the kind of bowel blockage that one cannot relieve. There was no surgery or panacea that could rectify the predicament. The end was nigh, and I knew it. The job was before me to let her know my thoughts. No matter how many times I am required to have the so-called "death talk," it is never easy. But in this case, I knew it would be monstrously hard. After all, she was more than a patient. She was a friend and an inspiration. Her husband had also become a good friend over the years. I knew they were both inside the room. I knew they were waiting for me. And I knew I had awful news. What I didn't know was whether or not I had the strength to let the words fall from my mouth.

I rarely cry in my life, but Jennifer got the best of me. I started by recounting all the places we had been over the many previous years, the reasons for the surgeries and all of the chemotherapies. I went over our successes and our setbacks. And then I led us up to that day.

"Jennifer, I'm going to tell you what you already know," I said. "I think you have known it for the last several days. And there is no simple way to say it." I held her hand. "There is a time to be born and there is a time to die. And death will come to us all. But this is your time to die and as much as I wish I could change that, I cannot. God alone holds that power, not me."

With that she began to let the tears flow, and feeling the emotion through her grip on me, I could not help but shed tears also. I told her I would help her through these days as best as I could. I let her know that even though hospice would be involved, I could help as she needed me to. And then I prayed for her. I quoted from Job when he said, "Yet though He would slay me, yet would I trust Him." After the prayer I just sat on her bed and held her hand while she and her husband cried. It was difficult to say goodbye to a friend, and she was a very good friend.

That was the last time I saw her alive.

I went to her funeral, which is rare for me. I have only been to the funerals of three patients. A piece of me dies with each of them, so I choose the funerals wisely. Her funeral was really a joyous celebration. They played a piece of music by Michael W. Smith called "This is Your Time." The words summed up her life: "This is your time, this is your dance. Live every moment and leave nothing to chance." The church was full of her friends and full of my patients. I had no idea how many of my patients were her friends. I hadn't been aware of how many of them she had touched. She had blessed so many lives. One of my patients was so inspired that she essentially started the South Carolina Ovarian Cancer Foundation. The organization has helped many of our efforts regarding this disease.

Jennifer's memories are still very real in my mind. I often hope another like her will enter our lives at the office. She was like no other. Often my staff and I will ask, "I wonder what Jennifer would have done in a situation like this?"

She lived every moment. She left nothing to chance.

Chapter 2

Size Matters

There was an odd evolution that happened to me as a gynecologic oncologist (GO), and it came about fairly slowly over time: I have a hidden desire to find the largest tumors on the face of the planet.

I'm not completely sure how this happened or why, but maybe it will make more sense if I briefly explain what a GO does. I take care of cancers that are unique to women. However, not every case GOs manage is necessarily a cancer. All ovarian cancers are considered tumors, but not all ovarian tumors are cancerous. Cancers do make up the bulk of my work and the work of most GOs, but some are just benign pelvic tumors. (There's also an in-between group known as "borderline tumors" or "low malignant potential tumors," which are malignant but rarely spread or take a patient's life.) Referrals for benign tumors come about because the tumor may seem technically difficult to extract from the pelvis by the referring general gynecologist. Since GOs spend much of their surgical life operating on cancers and/or difficult pelvic tumors, the referrals make sense. So half of what we do is surgical, and the other half is treatment planning and chemotherapy for women who indeed have gynecologic cancers. One of the privileges we have as GOs is that many of the most fascinating pelvic tumors are referred to us. Fascinating, you say? That's how I see it. Of course, I'm probably a little different.

You see, there are many tumor referrals sent to my partners and me, which we then remove. That's the point to my job. Now I know what you're thinking. "You get excited about tumor referrals?"

I have to confess, I do. It's a little bit like Forrest Gump's saying about a box of chocolates: You never know what you're gonna get. Most of the tumors are ovarian, though occasionally they're from the uterus or other organs. Of those from the ovaries, some are small and others are big--really big. So if I'm going to take tumors out anyway, why not take out big ones? It's a lot like a fisherman. He doesn't just want to catch fish. He wants to catch the biggest fish. Likewise, I want to take out the biggest tumors. I like them super-sized, like a Biggie Tumor with fries. For me, it's bragging rights. And even though I'm commenting about myself, trust me, I know your local GO probably thinks the same way. Shoot, I think they all want to have the world record. If I can't be known for being the fastest person alive or the smartest, why not be known as the dude who took out the huge tumor?

Most of the largest tumors in the world are ovarian by nature. I'm not totally sure why, but part of the reason is that the patient can't see an ovary, and therefore it can grow in relative silence. If symptoms are present, they tend to be vague. Also, as ovaries grow, they may not severely affect a person's quality of life. So if a patient simply did not want to seek medical care, or could not seek medical care, she may potentially ignore the growing problem and live (assuming the ovary is benign). If she ignores it, she may get away with it for a little while, but often in tumors such as these, they will keep growing. And growing. And growing. As incredible as it may seem, they can grow to be bigger than a beach ball.

Often the really big tumors present the hardest but best challenges from a surgical point of view, and most GOs thrive on that challenge. Here's a picture of the challenge. Try picking up a water balloon the size of a beach ball. It's not as easy as it looks. First off the beach ball is filled with a water-like solution, or liquefied fat, or perhaps a mucin-like jelly the consistency of molasses. At beach ball size, it weighs a lot. If you're still thinking, "No problem," consider this: When you pick up the water balloon, you can't pop it. It may be full of cancer cells. If you pop it, the cancer cells can spread everywhere. You don't have to be a doctor to know that spreading cancer cells everywhere is a bad idea. The pressure is on as you take the tumor out--this patient is trusting you. Remember, don't pop it. Good luck!

As a GO, I had set a goal that I wanted to achieve at some point in my life, and at the time I thought it was wishful thinking. This goal was based on a voluminous tumor that one of my friends and colleagues had removed. Among my close friends that do this for a profession, that record was 66 pounds. I realized, though, that it was unlikely I would ever attain this lofty goal. Cases like that don't come along every day.

There are many prerequisites that go into growing a tumor that large. One, the patient has to be physically able to handle a 66-pound tumor. Two, it can't be much of a cancer, or by the time it gets that big, it will have spread and taken the patient's life. Three, the patient has to be willing to totally ignore this thing for a long time. How many people can carry a tumor greater than 66 pounds and not want to have it removed? Four, the tumor has to be one that is growing at a reasonable rate. I could probably think of about 30 other prerequisites, but the point is, a lot goes into making something this phenomenal. The stars must completely align. Oh, I had had one or two stars align a few times in my career. I had a couple of measly 30-pounders. Not the kind you write home about. A tumor that size is impressive, and you may get a few ohhs and ahhs, but it's not unbelievable.

This, however, is the story of how the stars did align for me one day.

I normally operate on Mondays and Thursdays. The other days of the week I see patients in the office. Occasionally, I will operate on office days for cases that can't wait, or to catch up if the surgical schedule gets behind. But the perfect storm occurred on an odd week. My partner was out of town. There were two of us at that time in the practice that took care of the gynecologic oncology in this community of around 500,000 people. When one of us was out of town, it usually meant the other one often did surgery on non-surgery days. I'll never forget that I had posted two cases on a Friday, which was highly unusual, and both cases were ovarian tumors. One of them came in at six pounds, the other 20 pounds. It was an odd preamble to what ultimately transpired.

My office called me during the second case to tell me a physician about an hour north of us needed to talk with me pretty urgently about a case. Those types of calls come in often. I finished up my last surgery and took the call.

"I have a case I would like to discuss with you," the physician said. "I have a 35-year-old white female that presented to the emergency department. When she came in, the ER doc thought that she was in liver failure. The reason they thought that was because her abdomen was huge. It looked like she had massive ascites."

Let me help you with a little medical terminology. Patients that end up in liver failure often get a very protuberant abdomen. This is from fluid (ascites) that fills up the belly. The patient, in every sense, looks pregnant. Women with ovarian cancer also may look the same way. Often ovarian cancer patients have a belly full of fluid. So, ascites is often the end result of both liver problems and ovarian cancer. They are caused by different mechanism, but in the end the look is the same.

"Interestingly," he continued, "she had normal liver enzymes. Given that the ER doc thought that she wasn't in liver failure, he then ordered a CT scan. When the scan came back, it became apparent that the fluid was not freely in her abdomen. In fact she had no ascites at all. But the protuberance in her abdomen was a tumor. And let me just say, it is a very big tumor. The biggest one I've ever seen by far."

He had my attention. I salivate over opportunities like these. I asked him a few other medical questions about the case to clarify the picture for myself. Once I had completed my inquiry, I was somewhat satisfied that the history was indeed consistent with an ovarian tumor and/or an ovarian cancer.

Before he hung up the phone, he repeated one more time, "This tumor is very big. I mean really big."

I worked out the transfer, and she was on her way. All that was required now was a little waiting. Even though it was Friday afternoon, I decided to wait around until she arrived. I was like a kid waiting for a Christmas present. What he described made me think that this could be the record. A personal best. The perfect storm. Had my ship come in? It was worth the wait to me.

About two hours later, I got the call that she had arrived at the hospital. Her CT scans had been sent with her, and I immediately loaded them in the computer, even before I saw her. What I saw was breathtaking. Her abdomen was about 90 percent tumor and 10 percent non-tumor. Her insides were squashed in her belly. I was surprised by the films alone. I could not believe she could eat or breathe, much less that her kidneys had survived this long. I couldn't even imagine that she was alive. I had heard about this kind of case but never experienced it.

I got a nurse and walked in. The picture was striking. The first thing that struck me was the overwhelming size of her abdomen. Under her covers was what appeared to be about a 250-week pregnancy, or a normal full-term pregnancy with at least 10 children inside. I also noticed her emaciated shoulders. She had prominent bones and no real muscle mass. It was a picture of a malnourished woman with a huge belly. She was overweight on paper, but in reality she was actually quite underweight. Without the tumor, I could tell there would not be much left. Another striking feature was that when she smiled, she only had about four teeth, and they weren't exactly healthy. But she did smile. Whatever she had been through, she still had a sense of humor.

I introduced myself to her and asked her some questions. After a series of basic background questions to understand her medical history, I proceeded with the questions about the mass.

"I have to ask you, how long have you had that?" I asked. I would never laugh at a patient, but I did have to smile when I asked her that. The mass was so obvious, you couldn't ignore it. It was the elephant in the room. We had to talk about it.

She smiled back and said, "At least two years."

Of course, there were thousands of questions that then came to mind. How could anyone live with that for two years? How could you walk? How could you sleep? How could you turn over? For a minute I just couldn't find the words I wanted to say. Above all, I wanted to be kind and guard my words. And yet I had an inquiring mind. How? Why? I gathered my thoughts.

"Do you mind if I ask you one more question?"

"Not at all," she responded.

"Why now? I mean, if it's been there for over two years, why would you come in now?"

"I take care of myself and I love to garden, and it got to where I simply couldn't do it anymore," she said. "My quality of life was so bad that I decided I simply couldn't live like this anymore. It was time to come in. So here I am. Can you fix it?"

"I'll do my best."

Before I finished, I had to do an examination. I'll leave out the details of the exam with the exception of one. The most amazing part of the physical examination was when she stood up. She literally had to lift up her belly. She placed her hands under the lower part of her stomach area to carry this mass. It was just about the most incredible sight I had ever witnessed. An otherwise thin woman who weighed more than 300 pounds, and it was all right there in her abdomen. Truly, it was a sight to behold. I had wanted a challenge, and the challenge of my career was staring me down. It was as if the tumor was saying, "Come get me out if you can." I had my doubts, but I would try.

I left her room and began to get everything in place to do her surgery. I must admit I was dumbfounded. My mind kept going back to the images you would have to see to believe. I would occasionally chuckle in my heart as I walked down the hall just thinking about this scenario. It was going to take a few days to get the team together and have the patient ready for the operation. I would need an assistant to help. I knew for a fact I couldn't lift the tumor by myself and operate underneath it all at the same time. I would have needed to put on my third and fourth arms to pull that off. Also, I had to devise a way to take the mass out without rupturing it. (Remember, the water balloon may be full of cancer cells, and a big tumor like this can be tricky.) Additionally, I needed to get the folks in anesthesia to see her about potential problems we could encounter in putting her to sleep and positioning her. Finally, I had to consider what it would be like to remove almost half of a person's body in a few hours. They don't teach you that one in medical school, residency, or fellowship. There are some things you have to make up as you go along. This was an adventure in uncharted waters.

Show time. I had my team together, anesthesia was ready and we were off to the races. Anesthesia began to put her to sleep. We were a little worried that once she was asleep the weight and size of the tumor would put too much pressure on the blood vessels bringing blood to and from her legs, so we had to position her accordingly.

The next challenge was getting her prepped for the surgery. She was all belly, and all of that belly would be involved in the operation. Every square inch of her belly wall was being pushed by this magnificent tumor. The incision was a cool 36 inches in length. (Yes, we actually measured it out for interest's sake.) The tough part was making sure not to cut into the mass as we opened the abdominal wall. This worried me because a mass of this size was pressed firmly against the anterior wall of the abdomen. There was not even a millimeter between the tumor wall and the opening lining of the abdomen called the peritoneum. I had to cut one cell layer at a time. That's not easy since the human eye cannot see a cell without a microscope. So I tried my best to keep the sharp knife blade from going through the tumor wall.

Fortunately, all was going well. I was in the belly, the patient was breathing, her blood pressure was stable, the tumor was intact, and her legs seemed to have good blood flow. Where to now? Staring me in the face was something of such gargantuan proportions that I was awestruck. The plan was to operate underneath this ovary without popping it, which would require a way of holding the ovary with evenly distributed pressure around it. So I had my nursing staff sterilize a sheet that I could wrap around the tumor. This allowed my assistant to hold the ovary up onto her pelvic bone so I could get underneath what seemed like a boulder. The sheet allowed for even distribution of pressure, and the pelvic bone offered a firm foundation to support the ovary. There was no way anyone could hold this mass up for as long as it would take to cut off all of its blood supply, so we needed that temporary resting place.

The blood supply came in on the underside of the ovary, where it was attached to the pelvis. The blood vessels that feed an ovary are normally about as big as a pencil, but these vessels were as big around as a tangerine. Normally I would just take one surgical instrument and clamp off the blood supply. But not that day. I had to dissect each and every vessel. It probably took five separate clamps to successfully and safely tie off all the blood supply. It takes a lot of blood to feed something the size of Canada.

After clamping, the tumor was free. It was lifeless. It lay there in the sheet, vanquished. It would no longer take from this poor woman, who was doing quite well. I then rolled the ovary back into the belly to completely surround it with the sheet so that lifting it up and out would be a safe process. It went as planned. I hoisted it out.

The tumor came in at 130 pounds! That's like having a full-grown person in your abdomen. Assuming the average baby weighs eight pounds, the mass was the equivalent of 13 fully matured babies. Carrying full-term twins would be child's play next to this. When the patient left the hospital, she went home at 142 pounds. She had lost almost half her weight.

The night of the surgery, I placed her in the ICU because I did not know how the human body would react to losing that much weight. I never had a class in medical school titled "Women who lose half their weight in one hour, and how to manage them." Fortunately, she sailed through her first night after surgery. It's amazing that the human body can take a situation like that and sort it all out. In fact, I would go so far as to say the recovery in this case was almost as easy as a straightforward hysterectomy. The biggest difference I noted was the size of the panniculus. When people have eaten just a few too many French fries and the belly hangs down over the pelvis, we call that a panniculus. The patient had a panniculus prior to surgery, but afterwards, her skin hung down almost to her knees. What was most impressive, however, was that in six months, the panniculus went from just above her knees to about two inches below the pelvic bone. I had never seen the skin and muscle retract that much that quickly, but it did. The human body is truly amazing!

In the end, the patient survived the operation, her skin retracted well, and the mass was a borderline tumor. She has been "cured" more than 10 years. The last time I saw her, she had those four dark and dying teeth replaced with gorgeous white dentures. Her smile was beautiful. She was back in the garden.

Plus, I now have the record among my friends, and it will be hard to top. Size does matter!

Chapter 3

In Harm's Way

Have you ever wondered how medical school is arranged? It's a pretty simple concept, but the devil is in the details. The first year is designed to teach students everything about how the human body is supposed to work. That includes classes on topics such as anatomy (where everything is and what it's called) and physiology (how it works). It's obviously more complicated than that, and every medical school is a little different, but most of them tend to follow a similar curriculum. The second year involves learning about everything that can go wrong, such as diseases of the heart and liver and also disease processes like cancers and infections. Those two years are the two hardest years of school you can imagine. The amount of material you're presented with is overwhelming. There are many days you feel like you'll never achieve mastery over the subject material. But somehow, one day, you finish those first two years. The third year must be easier, right?

Well, it's not easier, but it's not necessarily harder, either. It's different. First of all, instead of spending every waking hour in the library, more time is spent on the wards. And because of that, this third year doesn't take place just during daylight hours. It's an around-the-clock investment. I stayed up all night a few times in high school for fun, repeated it a few times in college when I got behind in my studies, and a few more times in early medical school in order to stuff last-minute facts into my brain. But in that third year, staying up all night was a requirement. Sleep was something I wanted most but rarely got.

When I started my third year of medical school, my first rotation was on a surgical service. This is where I learned that sleep was for the weak. The quicker I got used to the notion that sleep wouldn't happen, the quicker I became fully acclimated. I somewhat recall the first night I came home from call. I had just invested 36 straight hours of my life in taking care of patients who had been entrusted to me. But other than sitting down with my wife to eat dinner, I remember nothing. She said I basically fell into my plate. I wouldn't remember, since I was asleep in my lasagna.

It was a fun time, believe it or not. I was excited beyond belief and would have to pinch myself to make sure the experience was actually happening. For the past six years, I had dreamed of "practicing" medicine. And I was finally doing it. Well, kind of. I did lots of what they call "scut" work. I started IVs. I drew blood from patients and collected urine samples. But did I save lives? That would probably be a stretch. Still, I was indeed in the bowels of Parkland Hospital in Dallas, taking care of patients. And that was cool.

On this first surgical rotation, I was assigned to the vascular service, which is concerned with things that go wrong inside blood vessels. My life consisted of helping with amputations of limbs, cleaning out vessels (the Roto-Rooter service), and fixing or bypassing weak and diseased blood vessels. I saw more than my fair share of blood. I became an expert at holding retractors for other surgeons so they could see where on earth they were operating. I also learned to develop thick skin, because I got yelled at on occasion for not holding the retractor correctly. To be fair, I was about two feet from the table and couldn't even see what I was doing. The doctors placed me in contorted positions and asked me to stand that way for about two hours. My arms would be outstretched, holding an impossible position that could not be maintained by any stretch of the imagination, for hours on end. Then when blood started to flow from the incision, it was always my fault, according to the attending surgeon. So thick skin was a necessary thing to develop.

At night, my job changed from retractor holder to errand boy. I would do anything or go anywhere I was asked and didn't ask why. It was my lot in life as a third-year student. But all the weird and wild things happened at night. It was a veritable smorgasbord of humanity that needed my attention and the attention of all my classmates. This is when the "knife and gun club" and all the looney tunes came out to play. They could get really weird, almost spooky. The alcoholics loved to ride their motorcycles at the witching hour and flirt with death. Drugs would flow into people's veins and do bizarre things to their minds. Night was truly the most interesting time of all.

One particular night was especially interesting. Just after starting my new job as a third-year errand boy, my chief resident, whom I followed around like a puppy dog, was called to the ER. He told me to follow him down, so I said "Yes sir," and off we went. There was a man there who had just been in a major car wreck. I figured it was another guy who had guzzled a few too many and wrecked his new car. When I arrived with my chief, that was far from the truth. This guy had gone from 60 to zero mph in about one second. He had literally hit a brick wall. I don't know why he hit the brick wall, he just did. In medical school we had studied this type of wreck, which is called a deceleration collision. That's when you go from moving very fast to very slow in a very short time span. It's a type of wreck known for tearing blood vessels in two, and as I said, my day job was helping on the blood vessel service. My team was quite adept at fixing broken and abused blood vessels.

When we arrived at the ER, the on-call emergency doctor pointed us to the patient. I watched my chief begin to interview him. I could tell my chief was watching this young man intently, studying his face to try to read between the lines. It was almost as if he suspected something was about to happen, like watching your fishing line when you know there's about to be a strike. And then it hit. In mid-sentence, the injured man just quit talking and slumped over in his chair. He was talking one second, and the next second he wasn't. It was apparent, even to my untrained eyes, that something awful had just happened. The patient suddenly had an ashen color. In seconds there were orders flying and people coming out of the woodwork to assist. All chaos had broken lose. The guy's blood pressure plummeted, and then it was non-existent. It took my chief about two seconds to recognize that the man had likely just ruptured a blood vessel, and a very big one at that. In the end, it turned out to be the aorta, which is the largest artery in the body.

The ER team was quickly debating all of the options before us. I couldn't really hear all the discussion, but the conversation was in high gear. Since I was the low man on the team, I was relegated to the back row. I wouldn't have known what to do anyway. It was scary to watch this unfold before me, but it was also exciting. I didn't know what the outcome would be or where this whole scenario was going.

Next, the guy was cut open right there in the emergency department. We weren't in the operating room because there was no time to get there. It was not a sterile incision. There was no time for that. If we had washed our hands, he would have been dead by the time we returned. Sometimes speed takes priority over sterility. Blood was everywhere. People were everywhere. Orders were everywhere. They were slamming in IVs and calling for blood from the blood bank. They were yelling for anesthesia just in case this poor guy woke up. After all, he didn't have any anesthesia or pain medications. If he were to have woken up, he would have definitely felt this one. He was split wide open.

This guy's aorta had a huge tear in it, and blood was pouring out everywhere. I had never seen this much blood at one time in all my life. I was watching a person bleed to death, and I felt like there was nothing I could do. When I did anatomy in medical school, the cadavers were dead and no blood flowed. When we were in the operating room, the blood flow was reasonably controlled because it was an elective procedure. But here, for a tense 10 minutes, everything that could be done was being done to stop the bleeding. There was suturing. There was pressure being applied to the bleeding blood vessel. But in the end, it was to no avail. The young man died right in front of my eyes. That was also a first for me.

One minute we're talking to the patient, the next minute he's dead. I quickly saw why the third year is a different kind of learning. No one can teach you how to be prepared for this experience. Things like this change you. You realize your own frailty and mortality. It could have been me in that car wreck.

But there was still more to learn this night. Soon my own life would be in harm's way.

At this point it was about 3 a.m. My chief said they were hopelessly behind in the surgery ER and asked if I would go down to help. So as they were cleaning up everything from the previous blood bath, I was more than happy to make my way out of the trauma ER and into the regular surgery ER. When I reported to duty, the house officer was glad to see me. He asked if I had ever sewn anyone up. That was a definite no, but I told him I was eager to learn. How cool would that be, putting stitches in someone? Can you imagine sewing someone like you would a shirt or some other piece of clothing? The thought was radical to me.

The emergency department was divided into multiple areas. There were separate ERs for psychiatric cases, obstetrical cases, and so on. I was now privy to the trauma unit and ready to experience a new area. The surgical unit was a zoo. There were people everywhere. They were sitting in the waiting room, standing in the halls, and sitting on the floor. As you walked by the sea of humanity, there were bandages covering up all kinds of bloody wounds just waiting to be repaired.

The surgery unit itself was shaped like an "M," with the main desk a bar across the top. The three legs were hallways with six to eight patient examination rooms on each. But because the entire world had shown up that night, the hallways were being used as well. It was the only other option.

When I arrived to see my first surgical patient, he was on a gurney in the hallway. I introduced myself to the afflicted young man, and the resident began to walk me through the problem. Essentially, the guy had been out playing with the knife and gun club and ended up on the wrong side of the knife. On his forearm was a gash about four inches in length and not deep enough to have punctured anything too vital--nothing compared to the poor guy from the trauma unit. This looked like a job for a pitiful third-year student. So my resident began to teach me what to do. The first part was cleaning the wound. This was somewhat painful for the patient, but a necessary evil. I scrubbed and cleaned, cleaned and scrubbed. I got it to sparkle as much as one can make a knife wound sparkle. I then covered the wound in some antiseptic solution and got ready to sew the laceration. Now, the first time you do something like this to another human, it's not as easy as it looks. My hands were sweating and shaking. Fortunately I had gloves on so the poor guy couldn't see my hands wet with sweat, and I tried to hide the shaking. After placing the local anesthetic around the wound, I was off to the races. I put the first stitch in and tied it down. Not so bad. My confidence went up, and I got a little more bold with the second stitch. Two down and about 30 to go. My somnolent state had been replaced with a feeling of exhilaration. I was operating at Parkland Hospital. Talk about a rush.

After about the fifth stitch or so, a blood-curdling scream came from the front desk area. (I was on the lower right leg of the M, so the scream came from the top in an area I couldn't see, but it wasn't more than 10 to 15 yards away.) My resident, the patient and I looked toward the end of our hallway. We had no idea what was happening. But before we knew it, that sea of humanity was coming towards us with all deliberate speed, like animals running from a fire in the woods. I didn't know what they were running from, but whatever it was, it seemed life threatening. They looked terrified. Then clarification came--there was no mistaking it: Someone had opened fire in the ER. The sound echoed so loudly it was deafening. We were panicked, and it appeared the mystery shooter was coming our way.

For half a second I didn't know what to do or where to go. I was charged with caring for this knife victim, but there was a tidal wave of people literally running for their lives coming straight toward us. We had to move or we would have been trampled to death. It was that simple. So I dropped the suturing device attached to the needle even though the poor patient still had a needle hanging out of his arm. We all ran. Chaos was everywhere. My resident and I jumped into a broom closet that was close by. It was tiny, but it was all that was left. We figured everyone else, including our patient, had gone into other big rooms and slammed the door. It was like a game of musical chairs. But in this real-life game, if you didn't find a room, you might be dead. So my resident and I settled into the broom closet, the awful odor of ammonia all around us. But the amenities didn't really matter at that moment.

Then, shots were fired right outside our door. It sounded like a canon. I prayed intently for liberation. We didn't know who the enemy was, what he wanted, or whom he intended to hurt. We just sat on the floor and pushed our feet against the door with all our might, breathing as quietly as we could so the shooter wouldn't know anyone was in the room. But it was quite evident he was right outside the door. Where was the cavalry? Didn't the police know he was here? Wasn't someone going to save us? For 10 minutes we didn't move or speak. Finally, there was some stirring outside our door, the kind of stirring that would suggest friendly forces in the hallway. With great trepidation we moved our feet, opened the door, and ventured outside the safe haven. There was a bullet hole in the door not far from where we were. I had no idea who it had been intended for. But we were alive.

Suddenly, paramedics came flying by with the shooter, blood pouring from a bullet wound in his neck. He had been taken down, but nothing was severed, so he kept his life that night. (If you're going to get shot in the neck, it helps to be shot in a major trauma center.) Believe it or not, one of my buddies, who was shot at by the crazy dude, helped save his life in the operating room. Something ironic about that, n'est ce pas?

Well, let's pick up the pieces. Before all this happened I was performing my first operation. But where was my patient? I walked all over the ER looking for the first surgical patient of my career. I opened every door, checked under every gurney, and looked in every broom closet. But he was gone. He had simply disappeared. I guess I can't blame him. It was a crazy night. But I've often wondered what happened to the suture hanging out of his arm. I also wondered what happened to the arm itself, since I was still about 20 sutures from the end of his gash. I guess he figured he'd take his chances somewhere else. So if you ever run into some guy with a needle and a suture hanging out of his arm, tell him I looked for him. I tried to get the suture finished and the instrument out of his arm. I just didn't know where he went.

And so ended the night from Mars where I was indeed in harm's way.

Chapter 4

Almost Cut Down by a Logger

In order to fully grasp this story, you'll need just a little background on cervical cancer. For years scientists suspected cervical cancer was caused by a sexually transmitted disease (STD), but isolating the organism had eluded them. Eventually they figured out that more than 99 percent of the cases are caused by a sexually transmitted virus called the Human Papilloma Virus (HPV). The virus is the most common STD in the United States. It's highly contagious, accounting for more than 500,000 new cases of cervical cancer in the world each year. Interestingly, cervical cancer is not a very common form of cancer in the United States compared with the rest of the world, most likely because we have an accessible, effective screening program that takes its form in the commonly performed pap smear. Although the pap is not a perfect test, it does appear to be highly effective in locating most cases of cervical cancer and precancerous lesions of the cervix.

This disease process also has an identifiable precancerous change. Why is that important? If a potential cancer has an identifiable precancerous change, and it is easily treatable, this can potentially prevent the occurrence of cancers. Unfortunately, almost half of the cancers of the cervix in the U.S. are found in women who don't get screened. Since they choose not to be screened or do not have access to the screening, it goes without saying that the precancerous lesions of the cervix in those women are not treated. If no treatment is done, it inevitably may lead to cancer in many of these women.

Each cancer has its own unique way in which it spreads. Some spread through the blood, some through lymph nodes, and some spread by growing directly into the structures next to the cancer. (This list is simplistic and doesn't encompass all the ways in which a cancer may move about the body or metastasize.) Cervical cancer classically spreads by two methods. Usually it moves into organs right next to where the cervix is located, like the bladder, vagina, and/or rectum, or it spreads by moving through the lymphatic system into lymph nodes. Again, it can move in other ways, but these are the two most common methods.

The virus is readily transmitted from partner to partner, and there are many strains of the virus that people can contract, though some are more virulent than others. For instance, two strains of the virus, HPV 16 and 18, account for about 70 percent of cervical cancers. Thus, certain individuals carry stronger and more powerful types of the virus than other people who are infected. If you follow the female partners of men whose previous spouse died of cervical cancer, the subsequent partners obtain cervical cancer far more commonly than the general population.

Now that you're practically a doctor, I can delve into the story.

When I was not far into my oncology fellowship (a physician's final years of training), I got a call from the emergency room describing a patient I needed to address. As a fellow, you are at the beck and call of the emergency room 24/7. Although today there is a maximum of 80 hours that one can work while training, that was not true for many generations of physicians and it was not true for myself. But I'm not complaining. In fact, the day I was offered my fellowship was truly one of the most exciting days of my life. I loved the work so much that it was not difficult to be there. And frankly, training institutions seem to get the most interesting and bizarre cases of all, probably because those who seek the least amount of care generally end up at these facilities. Life was never dull then, and because I still am involved in training others, life is not dull now.

(Warning: If you have a weak stomach, you might want to skip over the next few paragraphs.)

I walked into the emergency room and met for the first time a logger from eastern Kentucky and his wife. The physical contrast between the two people was immense. He was around 6' 7" and had a beard down close to his navel. He was not overweight, but at that height he was probably on the order of 220-plus pounds. His clothing fit his profession. He had rugged outdoor boots, jeans, and a work shirt that had been through a few trees. This man was carrying his wife who, I'll never forget, weighed in at 67 pounds. When I saw this picture, I knew something was terribly wrong. She was so weak she almost appeared dead. Truly, she was like a lifeless doll. She did not pass the eyeball test, meaning, I didn't know initially if she would even make it out of the emergency room.

Another striking feature was the overwhelming smell. You would have had to be missing a nose not to notice. You don't think these sorts of things will affect you, but it is amazing how it can turn your stomach. The sense of smell is probably the least appreciated sense, but it is indeed powerful. There is a smell that comes with infections and the like that I don't even want to dwell on.

The husband began to recount a story, saying that his wife had gotten an infection, and they tried to clean the area and change her diet to help clear it. It was obviously to no avail. She drank lots of carrot juice in hopes that it would heal the process. This didn't work, either. Ultimately she got so weak she couldn't walk, gave up eating, and finally he brought her in. It made me want to cry. The power of denial here was profound.

After I got the history from the patient, then came an excruciatingly painful exam, not just from her point of view but also mine. The picture of what came next was indeed worth 1,000 words. There was no glossing over this one. It was neglect. Neglect so bad you didn't know whether to cry or get mad. It was beyond description. (If you haven't exited yet and are sketchy about your ability to hold down your dinner, this is likely the point of no return.) I went to examine her, and her bottom area was completely gone. There was no recognizable opening to the bladder, the rectum, or the vagina. It was just gone, probably for many months. I will spare you the detailed description, but one can imagine what was on this area. Now, I have a reasonably strong stomach and am almost never sick, but this case pushed the limit. I simply had to leave and regroup.

We admitted her and began to feed her through an IV to try to build her up. As soon as she was able, we took her to the operating room and made it so that her bowel and bladder stream did not come out onto her bottom area any more. This gave us the chance to clean things up, get a diagnosis, and figure out what indeed was the problem. We discovered that she had the worst case of cervical cancer imaginable. I don't know how many cervical cancers I have treated at this point in my life, but it has likely been more than a thousand. This was the worst.

The next several months resulted in chemotherapy and radiation treatment that would just bore you. In this setting, we tried to do our best to shrink the cancer, but realistically, I never anticipated that we could cure this woman. Ultimately, after months of care, the patient began to go downhill. This was no surprise to any of us involved in her care.

As is common with cervical cancer, when patients have an incurable situation, the pain that comes with the cancer is profound. Sometimes I spend more time with pain management than I actually spend taking care of the cancer. The pain is localized and extremely intense, and watching these women suffer is difficult. You want to help them, but sometimes the only way to help is to practically knock them out with narcotics. It's a constant balance between controlling the pain, yet not making the patient so sleepy from the pain medications that she can't even talk to her family.

"Mrs. Logger" was on mega-doses of narcotics. Her admissions were coming more and more often. It was a sad thing to watch. As the patient would leave the hospital, we would make sure that she was set on her pain medications and that the family was clear on how to have her take them.

One afternoon I got a call from "Mr Logger." The interesting thing was that he called me directly on my pager. I don't hand out pager numbers to just anyone. If you do, the little sleep you get as a fellow will be even less. Though as a fellow I was on call 24/7, the first calls went to my resident, who was in the hospital. This particular afternoon, we were within one or two hours of finishing in the operating room when I received the page, and I did not recognize the number. I thought it might be an attending at a number I didn't know, so I picked up the phone and made the call. He was calling to let me know that his wife was in a lot of pain, so much so that he couldn't stand to watch her suffer. (I never questioned from the day I met this couple that Mr. Logger loved his wife, even though what I originally saw would say otherwise. It was hard to know for sure, but it did appear he doted on his wife.) I suggested that we adjust her pain medicines to see if we could get her pain controlled.

He then let me know that she was hurting so much that she had taken all of her pain medications. As this conversation went on, his voice began to elevate as if to emphasize that the situation was my responsibility to fix. I let him know that I was more than willing to write for more pain medication or admit her acutely to control the pain. Neither plan was acceptable to him.

"If that's not acceptable, then what did you have in mind?" I said.

"You buy and provide the pain medicine, and I'm coming up to get it," he said.

I responded by saying that it was not my job to buy his wife's pain medicine, but I was happy to write the prescription. The conversation heated up fast.

"You will buy it and I'm coming to pick it up," he said.

I informed him one last time that I wouldn't provide the medication at my expense. He then told me that if I didn't have it in an hour, he would kill me.

"Kill me?" I thought. "What did this guy just say? This is what nightmares are made of. I'm just a poor fellow trying to survive the 100-hour work week and learn as much as I can. This is not what I had in mind for learning."

Now, I had been around this man enough to know that I might want to consider his threat seriously. But then I thought, "That's insane. He wouldn't really try to kill me." I sat there in a fog and contemplated this for a minute. Then I did what any good doctor would do: I got a consult. What kind of consult? A local one, to be sure. I was from Texas, and I was a long way from my stomping grounds. I figured that a doctor with a little more local flavor might be of benefit.

There was only one place to turn. Wade. He was definitely local flavor. He had the perfect Eastern Kentucky drawl and knew all of the regional colloquial terms. He was invaluable when it came to translating terms and phrases that I struggled with. It was like having a built-in translator. He was a chief resident and well versed in the goings on of the hospital environment and of Kentucky personalities.

"What do I do, Wade?" I asked.

"Buddy," he said with his perfect Eastern Kentucky accent, "call the cops. This guy will come after you, make no mistake."

I figured Wade understood the world we were in, so I called the cops. I felt a little uneasy making the request, but as I explained the situation, they decided to take it seriously. So picture this: A man is coming to murder me at the end of my day. It's not something that happens to you every day. In fact, I guess if the pursuer was successful, it would only happen once. I could see all those years of training go down the drain because at $2.75 an hour I couldn't afford to buy this lady's narcotic prescription. Mom never said it would be this way.

At the end of every workday, my clinical team would make evening rounds. There were usually about 10 to 15 patients in the hospital on the oncology service, and we would review all the patient items that would require attention overnight. As the senior fellow on the service, it was my responsibility to set all those plans into motion. It was also my duty to teach the group about subjects relevant to the oncology team. Normally we would have two students, three residents, and two fellows. On this particular night I was glad to be surrounded by the team, who also happened to be my friends.

But even more comforting were the two police officers accompanying me that night, both about 6' 5" and 250 pounds. They could pass for NFL lineman. Anyone who noticed the cops probably thought they were protecting some famous patient. Wrong. They were protecting a bottom-of-the-food-chain Gynecologic Oncology fellow. An average guy trying to eke out a living for himself and his family.

Guess what? Wade was right. The dude came.

When I saw him, I didn't know if he had a gun or if he planned to just take my head off with his bare hands. At that moment, these lineman in blue were my best friends. I felt like a cowardly quarterback dropping far back for a pass while the policemen were walking forward. Words started flying and the policemen quickly intimidated Mr. Logger into submission. Fortunately there were no arrests and no physical altercations.

I actually felt some compassion for the man. His wife was dying, and I'm sure on the inside he was dying, too. He obviously was not thinking clearly. That said, it's hard to have too much compassion for this man when his end game, as far as I knew, was to eradicate me from the face of the earth. I was hoping to be around to enjoy my wife's company a little longer.

But apparently my team had some compassion for the man, too. While the war of words was unfolding, unbeknownst to me, my team was taking up a collection to buy the morphine for his wife.

I never saw Mr. Logger again. It's hard to look often at men who at one point wanted to kill you. I'm not sure that he and I would have had a healthy future relationship. His wife died shortly after that, and it must have been an awful death. I can only imagine the pain of that horrible cancer. I didn't know a cancer like that could exist until I saw it with my own eyes.

But her death wasn't the only unsettling part of the situation. Remember, cervical cancer is caused by an STD, and this man was carrying the world's worst virus on his parts. That meant others may be vulnerable to these horrific strains.

About six months later I got a page from a number I vaguely recollected. I had seen it before, but I couldn't place it.

"Dr. Puls, this is Mr. Logger and I need you to talk to my friend."

Before his identity had even registered in my brain, he handed the phone off. After a moment I realized this was the guy who had wanted my head on a platter. All the memories were back, and it took my breath away. What on earth could this man or his friend want? A woman began speaking.

"Dr. Puls, I have started dating Mr. Logger, and he wants to know what I should use for birth control," she said.

All of my synapses were firing. I had just come off of seeing one of the world's worst cancers, which was caused by a collection of the worst Rambo viral subtypes ever collected. And yet here on the phone was a woman asking me about birth control with the man who could potentially cause that same cancer again. She didn't just need birth control. She needed to run very far away and very fast. The collection of STDs this guy had assembled had to have been some of the worst in history. I saw their damage lived out, and it was ugly.

I told her she ought to get some really good counsel before she considered anything further. But before I could say anything of substance the phone was dead. Was she a future statistic for another fellow? I hope not, but I can't control the world.

She was gone. He was gone. And so ended my brush with the worst cancer I ever saw and the virus that caused it.

Chapter 5

Beethoven

I learned to play the guitar in high school mainly because I loved the instrument and music, but also because it was the only instrument I could learn to play acceptably in short order. Initially I was into songs from the '60s and '70s. I liked pretty much anything from that time period that had acoustic guitar in it. There wasn't much breadth to my musical appreciation. I wasn't sure what existed outside of rock and folk.

    After high school, I didn't know what I was going to do with my life. Even though I loved music and would have loved to play with a group of musicians, I wasn't gifted enough to be in a band. You have to have talent for that, and I was simply average. So I wasn't sure what my calling was. I didn't think I was ready for college, although all of my friends were going. I had applied and gotten into the schools that I wanted; the problem was I didn't really want to go.

    I did have one idea, though. When I was in ninth grade we hosted a foreign exchange student from Brazil in our home for six months. It was fascinating to watch him go from not understanding our culture or language to adapting American habits and speaking English quite well. It intrigued me to the point of considering the foreign-exchange route myself. I talked it over with my parents and decided to go for it. I still can't believe that I talked myself into it and that my parents let me go.

    In the program I chose, the student did not get to choose the country. It was chosen based upon a self-description that was matched with a family in Europe. Ironically, I figured they would send me to Spain since I took Spanish in high school. That turned out to be a stupid assumption. It was based more on who I was and who the family was. So they sent me to, of all places, France. France! I did not have a clue about the language or the culture. It was hard to get excited. I mean, how much fun could France be? (Remember, I was 18.)

In the end, I was matched with a family I came to love very much. They were wonderful people. Yes, we had our growing pains and some disagreements, but ultimately it was one of the most stretching events in my life. When you take someone out of his culture and remove a common language and other points of reference, it has the potential both for disaster and growth. Fortunately, I grew, sometimes in ways I would have never thought of--like the 17 pounds I put on in one year. But that was a testament to great cooking and no discipline. I've never lost those 17 pounds and unfortunately have put on a few since then. (My wife says I needed them. I don't agree.)

This French family had a son, Antoine, who was three years younger than I. We got along marvelously. We connected on so many levels that even to this day, 30-plus years later, we both consider each other brothers. One of the joys he introduced me to was the world of classical music. He particularly loved Beethoven. He loved Beethoven's sonatas, piano concertos, and especially his nine symphonies. Shortly after I moved to France, Antoine bought the entire collection of Beethoven's nine symphonies on vinyl. He was so proud of this collection. He and I spent countless days listening to them, and, truth be told, talking about girls. If there were girl issues to be hashed out, he would walk into my room and announce that it was time to listen to Beethoven and play chess. That was our universal setup to discuss the fairer sex.

In the midst of these discussions, I came to love the final movement of the Ninth Symphony. It contains a choral portion that is in German, a language neither of us spoke. So one can imagine a French guy and an American guy, neither of whom could exactly sing, belting out the choral in a language neither one of us could speak. At times, I'm sure our neighbors, family and friends just wished we would go away. But the love of Beethoven's music was one of the many things I brought home from my year abroad.

Thirty-plus years later, I have a large collection of CDs and an iPod full of music. I never forgot my roots of learning music in the '70s, but I now have a fairly eclectic collection, with folk, rock, some country, and my classical collection, which is made up of predominately Beethoven. I guess one could say that I have tunnel vision when it comes to classical, but the music of Beethoven still has a deep meaning for me.

My circulating nurse of ten years used to keep my musical collection in her locker, and on the days that we would operate, she would bring the CDs into our operating room. Factoring in the difficulty of the case, the length of the case, the time of day, and the mood of the day, we would make our pick. Music can be a soothing force for everyone in the room.

Before one particular operation, though, I got quite a peculiar musical request. The patient was a nurse practitioner from another city who had previously come to see me about a mass that was growing on one of her ovaries. At one point there was some question as to whether this ovarian tumor was benign or malignant, but I believed we had a benign situation. When I told her that, she seemed relieved, as one would imagine. But then, out of nowhere, she asked if she could speak plainly with me about something. Without giving any details, she said she was very skeptical about physicians in general. That was odd and a little mysterious, but not a big deal. Something could have happened to her or a friend in the past, or she could have witnessed something as a nurse.

After telling me this, she said she wanted her surgery done through a laparoscopic approach, which involves making a small incision in the belly-button, plus a few other punctures, and retrieving the ovary through these small openings. The advantage to this approach is that the patient can generally go home the same day, and the recovery is shorter than traditional surgery. Her request was by no means unreasonable. Many patients make the same inquiry. I told her I would try to accommodate her request, though sometimes, for various reasons, patients will still require a straightforward incision.

So the surgery was set. We would go in with a laparoscope, look at the mass and see if we could get it out laparoscopically. She was down with the plan and so was I. She consented to opening the abdomen if the mass was malignant or if the surgery was too difficult. But we were to avoid opening her belly fully if at all possible.

When the day arrived, she was wheeled back to the operating room. Usually when this is done, the patient is given a little Versed, an amnestic drug that will typically put a person in la-la land. I've had two surgeries myself and have been in that la-la land. You quickly understand why the drugs they give are controlled substances. They make you feel way too good. They also make you way too free with your mouth. Things that you might never say in public may come out. It's a very vulnerable feeling.

So there we were in the OR suite, with the patient prostrate on the table. I was present to watch her through the induction of anesthesia. After the dose of Versed, she motioned that she wanted to say something.

"Dr. Puls, can I talk with you?" she asked.

"Sure," I said and walked over to her.

"Are you going to listen to music while you operate?"

"Normally I do, but if you don't want me to, I won't," I said.

At this point, anesthesia held off giving her any other medications so she and I could finish our conversation. That was fine with me, since I was anticipating a light-hearted discussion.

"What kind of music do you listen to?" she asked.

"Truly, I have everything from rock to classical."

"What classical music do you listen to?"

"Mainly Beethoven," I proudly announced.

She deliberated a moment before speaking again. I thought she was about to say something quite profound in her somnolent state.

"Do you have Beethoven's Sixth Symphony?"

Since I had never had a patient ask me about my CD collection, I had to think for a second. "You know, I think I have the First, Fourth, Fifth, and Ninth Symphonies," I said after a pause.

Most people don't know much about the symphonies or just don't care, so I began to think this young lady must be a connoisseur of fine music. Perhaps I was about to be learn something about the "Pastoral" symphony, something I had never realized or had overlooked. I love it when people teach me things like this.

"Is there something particular about the Sixth that you love?"

"Have you ever seen the movie *Soylent Green*?" she asked.

I didn't get the connection between the symphony and the movie. I laughed a little bit inside. I figured she was under the influence of Versed and couldn't keep her focus on the conversation. Little did I realize she was actually quite lucid.

"Do you mind if I tell you about the movie?" she asked.

"Sure, why not." I would never deprive a soon-to-be-sleeping person of speaking her mind.

She proceeded to tell me that this movie was set in a time when there were too many people in the world. (I'm not sure whether she had her facts about the movie straight or not--I've still never seen it--but the veracity of her description is irrelevant, since this is indeed what she believed to be true.) With all these people in the world, there needed to be some form of population control. So the government decided to kill people, grind them up, and turn them into food to feed the others. The name of the food was Soylent Green.

While describing this, the patient asked me if I knew how they killed them. So here I am about to operate on a woman with a cyst on her ovary. We had been discussing Beethoven, who is one of my heroes in life, and now all of a sudden she's telling me about a government that kills people and grinds them up into food. It was odd, and as you might imagine, I was getting uncomfortable. Remember, too, that she was skeptical of doctors. Anyway, she reveals that the victims in the movie were put to sleep with anesthesia while listening to Beethoven's Sixth Symphony.

I hesitated for about five seconds. I was really freaked and honestly felt paralyzed by the situation. So I simply turned to anesthesia and made a hand signal that it was time to go to sleep. This conversation was over. I needed to get the job done.

Sleep was induced and the laparoscope was placed in through her umbilicus. Probably with a cold sweat on my brow, I visualized the ovary, and did the things that I do and took it out. Praise the Lord, it went as planned and the entire procedure was done in 30 minutes with barely an incision.

Two weeks later the patient came to see me for the post-operative visit. I was pleased to see that she had recovered well from the operation. The pathology report was everything we had both hoped for. She was released to go back to work. She could not have been more gracious to me. She said that I had restored her faith in doctors keeping their word. Right as she was saying that, there was a knock on the door. My nurse popped in to say there was someone on the phone from the operating room. I asked to be excused and then went to take the call, expecting to be right back. Wrong. Someone was bleeding and needed help. While running down the hall, I asked my nurse to apologize to my patient for leaving prematurely. That was the last time I ever saw the nurse on whom I operated.

That night when I came back from surgery, only my regular nurse was left. As she was leaving she turned back and said as an afterthought, "That patient who was here earlier, you know, the nurse practitioner, she left you a present on your desk."

I went back to my office and on my desk was a letter and two boxes. The letter she had written was one of the most kind and gracious notes I had ever received. She reiterated some of things she said in the room earlier. Then I opened the first box--homemade cookies. And in the second box? A copy of the Sixth Symphony. Yes, Beethoven's Sixth symphony.

I don't listen to Beethoven's Sixth anymore. As big a fan as I am, it will never be the same for me. In fact, with the exception of the Ninth Symphony, I don't listen to much classical music at all anymore. She really rattled me. But I learned not to ask questions when patients are in that odd place somewhere between awake and asleep, and I certainly don't ask patients if they have any last-minute comments before going to sleep. I'm afraid of what they might say.

Chapter 6

A Tale of Two Husbands

"Till death do us part." How many of us have really thought through those words, even those of us who have recited them? There seems to be a lot lost in that recitation when so few marriages survive to the point of death. For many people, the words are empty.

It makes sense in some respects. When two people are standing at the altar reciting their vows, they're usually both in their prime. The man is young and strong and has his whole future in front of him. Perhaps his career is just taking off. He thinks he will control everything about his destiny. Truly, the world is his oyster. His beautiful bride is at his side, and in his eyes, she doesn't have a single flaw on her entire body. Perhaps she's thinking about her future family or her career. There's nothing but brightness on the horizon. So it's no wonder we take those vows, because we don't really believe that death will happen anytime soon. At that age, we don't comprehend the ramifications of aging bodies, and few of us have a real sense of inevitable death. Certainly, not many young people consider the possibility of a life-threatening illness like cancer.

But sometimes "till death do us part" hits couples squarely between the eyes, and they rise up and show us what the vows really mean. Sometimes marriages work so well they leave me in amazement. Two couples in particular made a lasting impression on me. It's the story of two women and the men who loved them unconditionally. These men watched their best friends suffer and die, yet they supported them, encouraged them and met all the challenges along the way.

Ironically, both of these women had the same first name. For the sake of our story, we'll call them "Jill" and "Jan". The parallels between the two were uncanny. Jill was around 45 when she was diagnosed, and Jan was about 40. Although their cancers were discovered about three years apart, they ultimately succumbed to ovarian cancer within months of each other. I'm not sure how well they knew each other, but I do know they became friends. Many of the women in our practice with ovarian cancer take their chemotherapy in the same suite and at the same time. Over the years, as one would surmise, they have the opportunity to get to know each other. I sometimes refer to the chemo suite as the "gossip suite", where everyone compares notes on any and every subject.

Jill battled her cancer for about 10 years and Jan fought hers for about seven years. Both of them were diagnosed with stage IIIC disease, meaning the cancer had spread from the ovaries to the upper abdomen by the time it was discovered. Jill previously had a cancer that started in the bone of her thigh, but this occurred when she was much younger and fortunately it was cured. Otherwise, Jill was in excellent health, which is probably why she was able to fight for so long. Jan was without any other medical problems when I met her. Just like in Jill's case, this allowed her to wage a strong battle for many years.

Jill's husband, "Chuck," was a photographer by trade. He and Jan had been married for many years and apparently had a wonderful marriage. He was very engaged not only with his wife's treatments, but also with the office staff. An oncology practice is a little different from some other medical practices because the patients are there so often. Chuck knew about everyone's kids, where they went on holidays, when they were moving, etc. He cared for the staff as a good friend would.

Chuck knew that I went to Texas every Thanksgiving for a week to be with family and to hunt. I'm not an avid hunter, but once a year I hunt quail and deer for just a few days. One of my partners, however, is a hunter extraordinaire. He has hunted all over the world and has the proof to show for it. I'm an amateur next to him. One year, Chuck asked how I had done on my Texas hunting trip. I told him it had been a great year--I had landed a 13-point buck with a reasonable spread in his antlers. He asked if he could see the picture. Once I showed it to him, he immediately asked if my world-class hunting partner had seen the picture. I told him that he knew of the deer but hadn't seen it. Chuck then asked if he could have the picture for 24 hours and assured me it would be worth the time. I knew he was an artist of sorts, so I gave it to him. The next day, a most satisfying picture showed up. A mere 13 points had gone to at least 35 points, and the spread on the rack had grown from around 26 inches to at least 50 inches. This picture immediately became my partner's Christmas present. I framed it and wrote the words, "Eat your heart out." To say the least, he drooled. Envy was the emotion of the moment. (I let him have these thoughts for at least 10 minutes until I could contain myself no more.)

Chuck also had the physical build of Santa Claus with the whitest of hair, and every year around early fall he would begin to grow an impressive beard. There was no mistaking that he had one of the most convincing Santa looks ever. I've never known anyone else to have a resemblance that striking. During the holidays, Chuck worked at one of the malls in the upstate area of South Carolina, and he also brought freshly made candy canes to the office every year for almost 10 years while dressed in perfect attire. The other patients loved to see Santa come around. I would joke with Jill that she had Santa right there in her home. She could ask for any gift she wanted. "Yes," she said, "and I get to sit on his lap every night."

Jan was married to "Drew." Drew must have been an athlete in high school. He was around 6' 3," very athletic looking, and just about the nicest guy. He was trying to hold down his job while supporting his wife in all she needed. They had no children at home and not a lot of support in town, as far as I could tell. The two of them had to tag-team everything. What stood out most about Drew was his willingness to go anywhere and do anything Jan required. One day she asked me what she could do since she was so tired. As I began to brainstorm about medications, activities and the like, she got an idea.

"I need a trip," she said.

I told her to talk to her husband. I didn't mind if she traveled. In fact, I think it's good for patients to travel so they realize they can fight cancer and still live their lives. Then she got a gleam in her eye.

"Could you make Drew take me somewhere?"

"I doubt it," I said, "but what do you have in mind?"

"Maybe you could write me a prescription to go to Cancun."

It would clear her mind, she said. It would get her to exercise. It would help keep up her nutrition. And a little sun would give her some color. Jan went on to explain how she had been through so much and undergone so many treatments that she needed a break. What do I say to that? It's hard to argue with a woman who has ovarian cancer *and* has convincing points. She should have been a professional negotiator.

So I did it. I wrote a prescription that said "One Trip to Cancun- no refills". And guess what? He took her to Cancun. From then on, every time she got overwhelmed with all she was going through, she would ask for a new prescription. She was like a drug addict, and the drug was a week in Cancun. After about the third trip, Drew came in to talk with me about how these "prescriptions" were setting him back. He jokingly asked me to go easy on them for the sake of his wallet. The two of them obviously loved that place because they must have taken seven or eight trips in the time that I knew them.

Drew came up with another diversion for the two of them (maybe to replace the trips to Cancun?). The two of them bought a Harley-Davidson motorcycle and transported it to places like Myrtle Beach. Then, of course, he had to buy her a whole new wardrobe of clothes to fit the hobby. She enjoyed the Harley for a while until one time, as she was sitting on the bike, Drew got up and it fell on her leg. I think that made her think twice about this adventure. Eventually, as Jan slowly began to lose the battle to cancer, they ended up selling the bike.

It wasn't all tropical vacations, bike rides and Santa costumes for these two couples. The progression of Jill's disease began to require chronic chemotherapy. She must have taken more than 300 trips to the chemo suite in those 10 years. Also, as is common with many ovarian cancer patients, she started to have intestinal issues. It often starts with indigestion, then constipation. Finally, food becomes more and more difficult to swallow and keep down. It's as if a vice grip is placed on the bowel and doesn't let go. Naturally, weight loss follows, setting off a plethora of other problems. When individuals are malnourished, their wounds don't heal well.

I mentioned that Jill had been treated for a cancer in her leg years before, and that treatment required radiation on the right upper thigh area. In light of her progressing ovarian cancer, the area over the radiation field on her leg began to break down. I sent her to the plastic surgeon who attempted to clean and debrid the area. But because of poor circulation and malnutrition, the wound never healed. It required constant attention and dressing changes, and it was in this setting that I began to appreciate even more the work that Chuck did for his wife. Though it may be hard to believe, family members of patients often will not lift a finger to help out the afflicted person. I once had a patient who needed some help with an injection at home and some strengthening exercises. When I walked into the room on the supposed day of discharge, there were no less than 20 family members in the room. I asked if someone would be willing to help out with this reasonably small task. Every single person in the room declined to help. They were all "too busy." I had to stop the discharge, and it took another several days to put her in permanent placement.

I don't know everything Chuck had to do for Jill, but here are just a few duties I'm confident he performed. Chuck had to keep an intense schedule for her. He had to help her with her dietary needs as she was not able to eat like the rest of us. I'm sure he had to watch what he ate in front of her. He probably sat at the table and "enjoyed" her diet with her. He had to arrange all of her medications, including nausea meds and fluids that had to be given through an IV. I know he came along to almost every visit over 10 years and played chauffeur. He had to witness some very painful exams, because for me to measure the size of the cancer, each visit required a lot of pushing and prodding. Then came all of the dressing changes and care that went into a non-healing wound. Not only was that arduous to perform, I'm sure it was not pretty to look at. Finally, he had to endure watching her die in the bed they shared as a couple all those years.

Drew was no less impressive and supportive. The first part of his job was keeping up with her desire to get away to Cancun. But when the going got tough, he rose to a whole different level. Jill also began to have bowel issues earlier in her course than Jan. After treating her for five years, I had to remove several pieces of intestine. Fortunately, it bought her some valuable time. But in the final year of her life, the cancer began to place the vice grip on her colon. Ultimately she developed a blockage that required a colostomy, which is when the bowel is diverted so that it comes out into a bag on the abdominal wall. It's never a pleasant thing, and taking care of it is a messy job. This is where I was really impressed by Drew. He spent hours helping her with the management of this colostomy. That is no small work. Only a person who truly cares for you would do it willingly and often. If the device that went around the bowel didn't fit well, he would work through it like an engineer. He was always brainstorming on how to improve it.

Like Chuck, Drew had to manage all of the medications. This can be a very involved process, particularly when people are on pain meds and get confused. He had to deal with her dietary issues as she slowly lost the ability to eat. This required a diet that involved special preparation and shopping. He came to hundreds of office visits and made sure she showed up on time to all of her chemotherapy treatments. Drew had to help with rehabilitation after each surgery, including a number of intestinal surgeries that require more time than average to recover from. Eventually, he and I had some very frank talks. He knew the end was coming soon. Drew could have been off doing other things and pursuing his own life, but he stayed by his wife's side, and I never heard him complain. In the end, like Chuck, Drew had to watch his young bride slowly pass from this life to the next in their wedding bed.

About six months after Jan died, my wife and I were downtown on a date. After dinner, we were walking along the street, and as we crossed, we happened to run into Drew. He was still grieving deeply for the loss of his wife. He recounted to my wife and I some of his thoughts since her passing, and in the midst of this, he broke down. I know that seeing me probably brought back a flood of memories. Perhaps some of them were good. Perhaps some of them weren't so good. I know that if I could have written his wife another pass to Cancun, he would have given his right arm for it. But I could not.

About three months after Jill died, Chuck came up to the office. I was operating that day and didn't get to see him, but it's probably just as well. My staff told me that he completely came undone. They all looked at him like family. They remembered the Santa suit. They remembered the candy canes he brought. They remembered the care he gave his wife--the dressing changes, the hand-holding during exams, and all the other expressions of love he gave. He was not even close to having recovered from the death of his wife. The grieving process was probably still in its infancy.

It would have been easy for these guys to walk away from the nightmares they experienced. From the outside, it might look like they went through pure misery. But I'm sure they wouldn't have traded it for anything. They had those final days in which they fulfilled their vows. Honor personified is a true joy to behold. So I tip my hat to them because they, without question, fulfilled the vows they had taken so many years before. The next time you go to a wedding and see people recite "Till death do us part," remember there really are those who mean it and fulfill it.

Chapter 7

A Resident Distress

I have been involved in the education of resident physicians my entire professional career. When I finished residency, I began teaching gynecology to newly graduated doctors. It's a privilege, because I have the opportunity to shape their practice habits, careers and their lives. I drill them on countless topics, from the mundane (how to hold a surgical instrument and tie a knot) to the technically difficult (how to safely control bleeding in a patient). I see this work as so much bigger than myself or my residents. It's about the ladies we take care of. If we don't do it well, they suffer. So if I can pass on what I've learned to this next generation, then perhaps it will help future patients.

Residents are required to learn clinical information and surgical techniques from multiple professors. Much of their knowledge comes from textbooks, but another piece of it comes from laying hands on a patient. The surgical aspect of training comes from time spent in the operating room, where we start with a young neophyte's first incision and work with the resident until he or she becomes a competent surgeon. That transition takes longer for some than for others. But for residents in Obstetrics and Gynecology (OB-GYN), we are given four years to complete their training.

During those four years, the residents are required to do a lot of obstetrical and gynecologic procedures such as delivering babies and doing hysterectomies. Another aspect of the clinical training is the care that goes into managing hormones, pre-term labor, birth control, etc. Not everything the residents are exposed to is necessarily work they will do in the future. For instance, most residents in obstetrics have to spend time in the neonatal intensive care unit (the nursery that cares for ultra-small babies). Those babies are ultimately cared for by neonatalogists, but it's good for an obstetrician (OB) to see what goes into this care. After all, the OB is the one who actually delivers the premature baby. If he or she sees why one-pound babies struggle in life, the OB will work extra hard to prevent pre-term births.

Gynecologic oncology (GO) is a portion of the training that OB-GYNs must go through to get their degree. Going through a portion of that training does not make them trained to do that sub-specialty--to actually be a GO requires three to four more years of training after residency. But it's an important part of the four years because much of the ICU care and technical surgery is taught on that service. So, as many of my residents see it, it's a necessary evil.

When you spend an inordinate amount of time with residents over four years, many of them become more than just students. A certain bond is created when you climb into the trenches with these residents. It's not uncommon for me to feel a sense of loss at the end of the four years, particularly for the residents who invested the most. They're like family. I love to see residents go out in my own community and begin to practice, and if they develop good reputations, that's like icing on the cake. (If they develop bad reputations, I deny that I know them.) It's also a blessing to operate with them after they finish their training, to see how their skills have matured.

In addition to training residents, there's another part of my job that most people probably wouldn't expect. Although I'm a GO, I'm occasionally asked to help OB-GYNs with difficult cases. Most of the time those cases are gynecologic cases, but sometimes I end up going back into the labor and delivery suites to help the obstetricians. It doesn't happen often, but when I'm called, it's usually bad. Maybe there's out-of-control bleeding or a hole in some structure that shouldn't have a hole. There are a plethora of reasons that women can bleed excessively during delivery. So generally, if the call comes, it's a distress call and the need to come quickly is paramount.

OB-GYNs and GOs (and all other physicians, as well) need to have a shared virtue: They must have the ability to leave their egos at the door. Egos will just get you into trouble. One of the characteristics I watch for in residents is their ability to recognize when they are in over their heads. Common sense will generally keep you from getting yourself in that position. In medicine, no matter how long doctors have been practicing, they will have to call for help along the way. Occasionally a case is so difficult you feel like you're drowning. It gets so tough you can't see things clearly. An ego would keep you from making that call for help. I've called a few vascular surgeons in my life when I was drowning; I'm not so proud that I won't admit that. If something is dicey, I want the right people on board to help. So, I check my ego at the door and call the right people. In the labor and delivery suite, sometimes the GO is that person to call. Not because we do obstetrics anymore, but because we do so much difficult pelvic surgery and are well-versed in cases that have the potential to lose blood. So on occasion I found myself roaming the halls of the labor rooms.

Which brings me to the story of one resident I remember well. "Beth" was not far from graduating and had been on my service essentially all of the allotted time over her four years. But one particular month was different from the others for two reasons. One, she was my chief resident managing my service that month. She was responsible for all of the patients that we had on the oncology service. Two, she was pregnant. In fact, she was in her last week or two of the gestation; her delivery was imminent. This was her first pregnancy, and up until this point everything had gone well. I make it a habit not to pry into the personal lives of my residents, but when the person across the operating table can't scoot up to the patient because of a protruding, kicking mass, it's something you end up talking about.

On one particular day, as I left to go home, I heard Beth was having some labor pains, but I didn't know if the pains were in fact real labor or so-called Braxton-Hicks contractions (false labor). I left, had dinner and went to bed around 10 o'clock. About 2 a.m. I got a direct call from labor and delivery. Most calls usually go through the paging operator. My residents, however, have my direct number in case they ever have a dire emergency and need to get directly in touch with me. So when the phone rang at two, I woke up from my somnolent state to hear the voice of one of my other chief residents ("John"). Before I delve into what he said, it must be stated that John was given to hyperbole from time to time. He had been known to tell me some histories that weren't exactly true. He loved to convince people of some incredible story, only to reveal that it was all a joke, or at least somewhat embellished. So when I heard his voice my first inclination was to doubt whatever he said. He told me that Beth, who had just delivered 30 minutes before, was bleeding heavily and might not make it. I paused for a second and contemplated the words. When you're in a dead sleep and someone calls to tell you something like that, it takes a moment to get the synapses firing. I didn't know what to think.

"Is this a joke?" I asked. Jokes at two in the morning are not exactly funny. And jokes on this subject about my chief resident are definitely not funny.

He quickly assured me it was not a joke. He began to lay out the story in some detail, and it sent me into a cold sweat. I don't like to hear stories like this about someone who had become like family to me. So I said I would quickly get dressed and head to the hospital, which was about 20 minutes from my house. I also I told him I would call him back and get the rest of the information while I was in route.

I ran to my closet, put on my scrubs and took off as quickly as possible. Fortunately, at this time of the morning there aren't many vehicles on the road. I called to get more details. Beth had delivered a healthy child without incident about 30 minutes before. But obstetrics can go south on you in 30 seconds. (That's why so many women lost their lives to childbirth in the 1800s and before.) And this one did go south unexpectedly. Just as the placenta was beginning to deliver, the uterus turned itself inside out; it's called a "uterine inversion," and it can be a life-threatening situation.

The uterus is a fascinating organ. It normally weighs about a fifth of a pound, but in a matter of months it can carry an 8-pound baby. By the end of the pregnancy it has more than its fair share of blood supply. This is, of course, because it is supplying all the nutritional needs for the baby. When you look at the blood vessels that go to the uterus before pregnancy and compare them to the blood vessels during pregnancy, it is night and day. What once were little arteries and veins become gargantuan arteries and veins.

The uterus is also interesting in another respect. It's an organ that is essentially a muscle. It has the ability to contract, which is how it pushes a baby out during labor. But these same contractions clamp down at the end of labor and stop the uterus from bleeding. If you've ever witnessed a labor you've probably seen the obstetrician massage the uterus after the labor to force it to contract down. If it doesn't contract down, significant bleeding will inevitably occur. Outside of trauma, there are not many examples in medicine where one can lose so much blood as in labor. So when Beth's uterus inverted, it lost the ability to contract down. Thus the flow of blood began and did not want to subside. The OBs who were there tried to re-invert the uterus, but that's not as easy as it sounds. This particular attempt was thought to be successful. They thought the persistent bleeding was from a tear in the cervix (mouth of the uterus) or from up inside the uterus.

On the phone, John explained what was being done at that moment to control the blood loss and the associated issues relating to her blood pressure. When people lose a lot of blood rapidly, their blood pressure will drop precipitously. This has implications as far as kidney function and the function of other vital organs. If there is not enough blood pressure to adequately supply the organs with blood and oxygen, they will shut down. So after hearing all this I made some decisions about things we needed in the room and steps we needed to take to gain control of the situation. John was working on all of these while I was on my way.

For me, the most important thing at that juncture was to drive fast. Coming from one who rarely drives fast, that was hard to do. I have been accused by my children on numerous occasions of driving like a grandpa. Since it was 2 a.m., and the only people out at that time are drunks and police, I was on the lookout for police. When I got within about one mile of the hospital I was doing about 60 mph in a 30 mph zone. As I came around one of the final curves, the flashing lights came on. Hosed! I had a feeling that at that time of night there was a high probability I would encounter the police. I just had hoped our encounter wouldn't be like this. So I began to slow down. I needed to tell him as soon as possible about the situation unfolding. But then I thought, what if he doesn't believe me? I could have him call the hospital. But Beth would just be bleeding all that time.

But just as he was coming up to me and I had slowed down, he took off and sped around me. I don't know if he got another call or what, but it was a miracle I needed. I slowed down for the last mile and made it to the parking garage, then ran to the changing room and Labor and Delivery. All said, it took me about 30 minutes from the phone call to arriving at the delivery room.

When I came in the room I immediately saw a lot of blood. Beth didn't look so good. She was pale. There were people everywhere. I became instantly concerned that Beth was in disseminated intravascular coagulopathy (DIC). DIC is a state in which one has bled significantly and has used up all the parts of the blood that help to stop bleeding. When that happens, no matter what you do, the patient continues to bleed. People die from DIC. I had already ordered up blood products on the way to the hospital to resolve that issue as quickly as possible. The only conundrum was that those products take awhile to come up from the blood bank. So I was still facing the bleeding issues on two possible fronts. Was she bleeding from the consequences of the inversion and/or was she bleeding simply because she could not clot her blood? I didn't really know. It could have been one of these or both of these. But I needed to figure it out quickly. By ordering the blood, at least I knew that I was trying to reverse some of the bleeding issues, and the products were indeed on the way. The next order of business was to look for the source of the bleeding.

At this point Beth was still awake and listening to me discuss her case. The difference between her and a normal patient was that she understood everything being discussed. I'm sure she didn't want to hear all that I had to say, but there was no way around it. Thankfully, she didn't try to give me advice, because I did not want her input. I wanted a clear mind and objective thoughts that could take the information before me and deal with it appropriately. You see, if a patient happens to be someone you know well, that person can influence your usual standard of care. That's why in many states it's against medical policy to write medications for family members. Because of that special relationship, a doctor might be persuaded to write for a medicine that he or she would not normally give. It's difficult to know in every situation who is appropriate to care for and who should probably be referred to someone else. If the patient is a close friend, I have to think about that situation long and hard. It can be difficult, but sometimes that duty falls to you anyway.

There were multiple sutures in Beth's "cervix", but she was still bleeding profoundly. Was this all DIC, was it from a laceration or was it from some other source? After looking at the anatomy for about three to four minutes, I still couldn't tell what was what. Every time I would look, the area would just fill back up with blood, obscuring the view. I knew one thing for sure: There had been a lot of blood lost, and it was still coming. Much more of a delay would be seriously detrimental to Beth. It was time to explore her surgically. Heroics were needed, because without them, her new baby might not have a mother.

Just after I made that call, Beth asked to talk to me. She looked me in the eye and begged me to save her uterus. I gave her my word I would try, but that my main goal was to give her child a mother. I was unwilling to promise that I would absolutely save her uterus. I reassured her that if it were possible, I would do every thing in my power to grant her wish.

So anesthesia put her off to sleep and we made the incision. Layer by layer we went into her. I was truly curious as to what we would find. There were no less than 10 people in the room who all wanted to know what the problem was. Would she start bleeding from all of the areas I was now cutting through? Remember, I didn't think she had much ability to stop any bleeding. And would I be able to grant her that wish? If I believed her uterus needed to come out, would I be willing to do it or would her words just before sedation persuade me to do otherwise?

Fortunately, it didn't take long to get into the abdomen and see the situation. From the moment I saw her uterus, I had the answer. The uterus was still inverted. It was inside out, which meant there was no way for it stop its own bleeding. That explained all the blood on the floor. It explained why every time I began to look at the anatomy, it filled it up with blood and blocked my view. I did not want to be too hopeful, because you never know the outcome until you have absolutely fixed the problem, but I did smile underneath my mask. I knew I had a very good chance of saving the uterus and an even higher chance of saving her life.

The top of the uterus wasn't visible. From my vantage point, the top was probably somewhere down in her vagina. I immediately thought that what had been sewn was probably the top of the uterus and not the cervix. I quickly decided to open the uterus like a cesarean section for several reasons. For one, I needed to actually see where the blood was coming from and remove any doubt in my mind. You don't want to make an assumption and miss another source of bleeding. For instance, she actually could have bleeding from the cervix, as well. I needed to know. It's never good to have to go back to the operating room a second time because you didn't get everything done the first time. Secondly, I wanted to make sure there was not any evidence of the placenta still in the uterus, which could also be causing the bleeding.

When I opened the uterus, to my answered prayers, there was no placenta left. And the sutures were all at the top of the uterus, which revealed that the top of the uterus had indeed been confused for the cervix. This meant it was very unlikely for Beth to have a torn cervix. The blood was most likely solely from the inverted uterus, and that could be rectified without a hysterectomy. So I re-inverted the uterus. The top was now where it was supposed to be, and this allowed the uterus to contract like it's supposed to. I then massaged it for about 30 minutes while all of the blood products were given. Once those were given, I felt like we would be out of the woods. And, praise the Lord, the bleeding did stop, her pressure stabilized and we closed her up.

Once the surgery is done, most people believe the work is done. But the work is far from over at this point. Beth still could have gone into DIC even after the abdomen was closed. If that were to happen, the bleeding would not only come from the uterus, but also from the fresh incision we had made. Plus, when the human body goes through a trauma like this, the other organs take a hit. The kidneys have to sort out all the new blood products. Her tissues had to heal. Also, if her uterus became soft and didn't stay contracted, it could bleed again. Many things could still go wrong. So I placed Beth in the ICU overnight and watched all of her blood parameters closely. They stayed stable. One of the joys of taking care of young people is that their bodies heal so fast and respond so well to corrective intervention. She was no exception.

As time passed from the surgery and Beth showed improvement, I began to breathe a sigh of relief. It makes me nervous to take such intense care of a friend. It wasn't what I wanted to do, but it was what I had to do. It's daunting to think that you can go from operating *with* a person one day to operating *on* the person the next. Overnight she had gone from running my service to being the sickest patient on my service. Our health is a gift, and you never know when it will be threatened. It can happen in a second. One moment Beth was celebrating the birth of her child, the next minute she was wondering if she would wake up from surgery.

Beth finished her training, went on to practice medicine and developed a great reputation. She also had another child, fortunately by a cesarean section that went smoothly. I told her that if she planned on having another baby by the vaginal route, I would be out of town for it. That delivery room was not an enviable place to be that day. But the events that transpired went our way:

The child got a mother, the mother got a child, my resident had her life, and by the grace of God, she had a uterus.

Chapter 8

She's Greek to Me

The office, at times, can seem like an endless stream of humanity. Cancer patients in particular face problems that are taxing on the human body and soul. There can be bowel blockages, infections, depression, hormonal changes, end-of-life decisions and so much more. Each new morning in the office brings more patients struggling, and they have to be tackled one at a time. But sometimes the patients' struggles can take their toll on a physician's mindset and his office schedule.

    Some days it seems like everyone is doing well and we can get through 20 patients in about three hours with no real delays. Other days everyone has problems, and it takes what feels like six hours to get through the same number of patients. I often tell patients the advantage of coming in early in the morning is that you can usually get in and out quickly. There haven't been enough accumulated health problems to back up the office. The downside to the early morning visit is that my hands are still freezing cold--the water coming out of the faucet is very cold that time of day. But they'll certainly wake you up. Later in the morning my hands are warmer from all the wear and tear, but heaven knows if we're on schedule anymore.

    There are lots of things that can throw a wrench in the schedule. Some of the issues are patient related, e.g. the traffic was bad or the alarm clock didn't go off. Some of the scheduling problems are doctor related. Perhaps there are calls from referring doctors who want to discuss a particular patient. But while the doctor's on the phone, he also wants to discuss last week's football game. And since he's referring to you on a regular basis, you don't want to cut him off. Other times the patients in the hospital are so sick they require a lot more time on rounds. And then, of course, patient sickness can get us behind. If a patient needs an admission to the hospital for illness, that can bog down the system. The patient ends up tying up an exam room while we wait for bed transport.

But the issue that throws a schedule off more than anything else is the operating room (OR). Many days a call comes from the OR from a physician with a fairly simple question, but other times a more complex scenario arises in which I have to run down to the OR. Even more time consuming, though, are difficult cases where the referring doctor is up to his elbows in blood or scarring and needs help. Those are the ones we dread the most. When that happens, I may be in the OR for two or more hours, and it throws the office into a frenzy. Remember, the patients in the office are there for a reason. Their problems don't go away just because I can't see them right then. They need to be rescheduled, yet the schedule is already full for several weeks in advance. This can really bog the system down for weeks into the future.

Still, calls like these can also be exhilarating because it's fun to have a challenge. Usually the scenario unfolding in the OR is not going well, and they want you to fix it. What guy doesn't want to fix a problem? It's how we're hard-wired. We want to fix things and we love the challenge.

On one particular day a friend and colleague called to say he had a woman open on the table with ovarian cancer. She was there to have an ovarian cyst removed, but when he opened the abdomen, it was apparent that the ovary was malignant. General gynecologists do not routinely take care of ovarian cancers, so he called to ask if I could come complete the operation.

There are a number of areas in the abdomen that have to be biopsied or removed to complete the operation for ovarian cancer. This is done to answer a very important question: Has the cancer spread? We take out a number of lymph nodes and other tissues that are pre-assigned in all ovarian cancer patients. Looking at those structures without removing them is considered very inadequate treatment. How we treat the patient from diagnosis is driven in many ways by this procedure, called a "staging operation." So if we get a call about a patient who is opened up and asleep with ovarian cancer, we make every effort to get to the OR while she is there to complete all of her staging with the appropriate biopsies.

So I shut down most of the patient visits for the morning. It made a mess of things, but when you weigh rescheduling those visits with taking a patient back to the operating room, the rescheduling is preferable.

When I arrived in the operating room, my colleague gave me the update on the situation. He thought this woman just had an ovarian cyst because her blood test for ovarian cancer had not been particularly high. But when he opened her, there were solid-like mushrooms growing off the surface of the ovary that looked malignant. When he sent the ovary to the pathologist, they called back with the obvious. It was malignant, and it appeared to have started in the ovary. So I got all those other biopsies from lymph nodes and elsewhere in the abdominal cavity. While we were in the midst of this, I asked the physician a little about the patient. He said she was a Greek woman in her mid 50s with an accent that made it slightly hard to understand her, but that she was delightful. Little did I know this delightful woman would soon become like a surrogate mother to me.

After we completed the operation I ran back to my office and let the primary doctor talk with the family. It's sometimes hard for the families to see their loved one go to sleep in the OR with one doctor calling the shots and then wake up with someone else taking over that role. I usually let the family have the news broken to them by the physician they have known and trusted. Then, once it has sunk in a little, I show up.

When I went by the next day, I was not prepared for the circus. The room was full of people, and they all wanted to meet me. It was like meeting a family I was marrying into. I had to make my introductions to everyone. I explained to them what we had done and what the tentative diagnosis was. I walked through everything we would do in the ensuing months.

The lady who had been on the operating room table the day before was exactly like the referring doc had described. "Mrs. Pappos" was middle aged, motherly, and had a distinct Greek accent. She reminded me of my best friend during college and medical school, who was Greek. He and I spent a lot of nights studying and lamenting over the volume of work we had. As my friendship grew with him, I came to know his family, and I realized how tight knit the Greek culture was. They all knew each other from certain communities, and they all seemed to grow up together. This point was reinforced as I went through residency--one of the other residents training with me was also Greek. From my relationship with these friends and now this new patient, I came to understand what it's like to be from such a connected culture. The few times I visited my friends and went to an exclusively Greek party, it seemed like everyone was related, and if you weren't related, you certainly were family friends. Everyone I was introduced to would say something like, "I'm [so and so], his cousin's aunt's niece." I would nod as if I understood the connection, but honestly I was lost.

Mrs. Pappos reminded me so much of the mothers of my friends, mothering me just as they had. One thing mothers commonly do well is cook, and Mrs. Pappos was no exception. In fact, she owned a restaurant that she continued to run all the way through treatment. The reputation of the restaurant was outstanding. Every time I drove by the parking lot was full. I was first introduced to her cooking through a meal at Christmas. She insisted on bringing a meal for the office every year at Christmas. The first time this happened I had no idea what was in store. The first year (of probably six or seven) she showed up with a carload of food. I hadn't heard of most of the food. There was baklava, which I knew, but there were also plates with stuffed grape leaves, spinach plates, and all kinds of meat. It was a spread fit for a king. Her food was so good and plentiful that we could eat it for a week in the office. To show you just how good a cook she was, President George W. Bush ended up eating at her restaurant one time during an election run in the area. And believe it or not, she decided to come see me that day for her appointment rather than stay to serve the President. I chastised her for the decision. How often does the President come to your place? But she thought I would be mad if she hadn't shown up. I told her that if the President ever comes again, she had my permission to not show up.

Mrs. Pappos had a big soft spot in her life. Since she was from the "Old Country," she wanted to go back every year. The only problem was that she was on chemotherapy. During treatment, patients get medicines every one to four weeks. But she wanted to go for three months every year. This created a conundrum. One of the options for continuing her therapy was for me to come to Greece with her. That invitation was extended on a yearly basis, and one of my great regrets is that I never took her up on it. She said my wife and I could stay at her house and enjoy the great Greek culture. She wanted to take us to the Greek beaches. A pet peeve of hers were the beaches of Europe. She was always sure to tell me that the beaches in Greece were not like the beaches of France where the women wore little to no clothing. It was clear in her mind that French beaches were inhabited by less civilized people.

What she didn't seem to understand was that I had several thousand patients who depended on me. I also didn't get three months of vacation in the summer. Like a mother talking to her son, she thought that I should always be there for her. Her husband would try to "entice" me by saying that we (meaning the guys only) could go out every night and get drunk. That didn't appeal to me, but he thought it was great incentive.

So the compromise was that we would figure out a way to give her chemotherapy in Greece every summer. Since I didn't speak any Greek, I would communicate through her to her physicians in Greece. She was very emphatic that I still decide her course of action. She did not want the doctors there making the decisions. I think this stemmed from the fact that I became like family to her. And who better than family to decide important things like the type and dosing of chemotherapy? That was her opinion, anyway. So for several months I would establish a communication between her, her Greek doctor and myself. Once it was all in place she would leave for about three months. As difficult as it was to do all of this, we pulled it off for six or seven years.

When Mrs. Pappos would return, she wanted to thank me for everything I had done, so she'd bring me presents from Greece. Now, I had not anticipated anything like this. The whole concept came as a complete surprise to me. The first year she returned, she came back to the office for a visit and exam. After all this was done, she said she had something for me. She began telling me about Greek history and ancient Greek gods. To be honest, Greek mythology was something I had very little interest in. "Why is she teaching me about Greek mythology," I asked myself? After a long soliloquy, she began talking about the Greek god of mumps. At this point, I was getting uncomfortable with the conversation and had no clue where she was going. Then she handed me a box and said she had something for me to remember her summer by. There in her hands was a small box that was about 6 inches square.

When I opened the box, it was the head of the Greek god of the mumps. The head was hideous. It was deformed and had the signs of scarring from the infectious disease. Now, I have the world's worst poker face; I can't seem to hide any emotion or expression. I didn't know what to say. I didn't want to be rude, of course. She meant it with the utmost of sincerity, but there was something lost in the translation. I tried to show thankfulness on my face, but I'm sure I failed. All I could do was muster a smile and say thank you.

The next year was a little different. This time she got into a similar conversation about things that are uniquely Greek. She then told me about some sort of Greek cookie. The way to make this cookie was something that her family had perfected. While she was in Greece she had convinced her family that they needed to make a lot of these cookies for me to eat when she got back. So on arrival, she gave me a sack full of cookies. They were delicious beyond description. Then she had one more surprise for me in the sack. There was a bottle of Ouzo at the bottom of the bag. Much to her dismay, I had no idea what Ouzo was. But when I pulled the bottle out it was apparent this was no ice tea. I think it had some sort of warning sign on it in Greek. It was obvious this was an adult drink. I thanked her as much as my poor poker face would allow, but I'm sure she could probably tell I had no concept of just how good this present was.

That night I went home and showed it to my wife. We tried a sip of it, but that was the only time. I think it burned a hole through my mouth. Clearing my nostrils was the only redeeming value I could see. I know there are those in the world who love this sort of beverage, but I have to admit I'm not one of them.

From conversations Mrs. Pappos and I had, she knew my real mother was a breast cancer patient. So often after our visit and exam she would ask about my mother. Then she'd grab my face and begin to offer advice.

"You are calling your mother, right?" she would say. "You know that you need to take care of her. If she ever needs anything, you go home to Dallas and take care of her. Do you understand what I am saying?"

It was paramount to her that I take care of my mother. In fact, I could tell that if I didn't live up to her expectations in this department, she would let me know. Now, step back a minute and take a look at the big picture. Here was a woman suffering from ovarian cancer, and all she wanted to know was that I was taking care of my mother. You could tell the concept was in her very fiber. Men needed to take care of their mothers, and there were no exceptions--certainly not her doctor. She would keep me very humble by reminding me what my primary responsibilities were in life. And lest I forgot, she would keep me in line. It was her job as my surrogate mother.

I got invited to all of Mrs. Pappos's family events such as weddings and other affairs. "No" was not the answer she wanted to hear when it came to these events. She applied the most pressure for the Baptism of her grandson. She talked with me about the impending event for many months. Even before he was born she would let me know that when that day arrived, I was expected to be there. What do you say when your "mother" tells you to be there?

I reluctantly agreed to come. She said it would be on a Wednesday night because the Greek Orthodox Church was booked so far in advance and the family wanted to make sure the grandson was baptized. They couldn't get it done on a Sunday. She also told me that there would be a reception afterwards in the fellowship hall. In my mind, a reception is a short family visit with cookies and punch. So my wife and I showed up on Wednesday evening at 6 o'clock. There was one problem, though. We were the only ones there. I thought that maybe we had the time wrong or perhaps even the date, so I asked someone standing around and they assured me I was there at the right time and place. But I had forgotten that this was on Greek time. I should have remembered from my college and residency buddies that this was the Greek way of life. Parties always started late and ran late. How could I forget?

The baptism started about an hour late and nobody seemed to mind. Everyone else came late as if it was expected. I was still "at work" technically, because I was on call and had to go back to the hospital after the reception to finish my work. So when this affair started an hour late, I was not happy about it. But there was nothing I could do about it. We stayed for the baptism, which was actually quite intriguing. Once it was over I ran into Mrs. Pappos to let her know that I had come. She insisted that I come to the reception. She walked us over to the reception hall filled with about 100 tables. On the tables were the finest settings of silver and china, multiple plates and many bottles of wine. Not exactly punch and cookies. Again, I should have known. She could not do anything related to cooking in a simplistic way. That was not her style. When it came to food and celebration, she did it with pizzazz. That was her way.

Mrs. Pappos was a special woman. She loved to cook and entertain. She cared for all those around her. She gave me more advice than you could shake a stick at. She told me she loved me like a son on numerous occasions. As she was losing strength and still taking chemo, she would report back to the restaurant and substitute when help was needed. She was the mortar that kept her family together. But in the end, like so many ovarian cancer patients, her death was cruel. She began to have more and more bowel issues. She got to where she could not eat solids, and eventually she couldn't even keep liquids down. Her husband begged me to do anything that I could to keep her alive longer. I tried every trick in the book, but finally the cancer won.

Mrs. Pappos inspired me on so many fronts. She would remind me often to care for my children, to love my wife, and to call my mother every week. If I didn't do it, she would let me know of her displeasure. She taught me about family and the need to celebrate the events of life. She taught me a work ethic that was unparalleled, and I never wanted to disappoint her.

I hated the fact that she lost her ability to eat. I hate that for everyone who gets ovarian cancer, but it was particularly cruel in this instance. To her, the dinner table was a celebration. It was her livelihood. It was where all the great moments took place. It was where families solved the problems of the world. And it was where the truly great moments of life were spent. But in all of this, she never complained. She would just ask about me, in a way that only a mother can do.

Chapter 9

Hallowed Halls

If hospital hallways could talk, they would tell stories that no one would believe. The stories weave a tale of humanity at its best and its worst. Those hallways have made me laugh and cry, and they have certainly kept life from being anything but boring.

And so, I thought a series of short stories about life on the hospital wards might give those on the outside a small glimpse of that which takes place within. Perhaps the next time you visit a friend in the hospital, you may find yourself more aware of similar stories going on all around you.

When I decided I wanted to go to medical school, I thought it would be reasonable to spend a summer working in a hospital to see what it was like. After all, if I passed out or tossed my cookies every time I saw something a little disgusting, then maybe a different calling would be in order. So I signed up at a large hospital in Dallas, Texas, during my summer break. They had me do everything from bedpans and enemas to transporting patients around the hospital. Some of the tasks given to me did, in fact, take my breath away. But fortunately no one can ever say I passed out or ran out of a room with nausea. One of the highlights of my summer was going to the operating room to watch a surgery on a patient with whom I was assisting. It was, interestingly, a cancer surgery, which is what I ultimately pursued as my profession.

But of all the jobs given to me during that summer stay, none of them stand out like the man with delirium tremens (DTs), which is essentially the aftermath of drinking adult beverages by the bucket. This guy had been on an alcohol binge for a number of days. He drank himself silly. When people do this for an extended period of time, their bodies get so used to the alcohol that when they withdraw, they go a little crazy and hallucinate (which is why the DTs is also known as "the horrors" or "the fear"). First off, I didn't really know that people drank like that, although I did see a few guys in college who were probably close. Secondly, having never seen the DTs before, it was a little scary. Even though I was pre-med, and my friends figured I knew a lot about medicine, I essentially knew nothing. The extent of my knowledge was understanding a mitochondria or where to find a nucleus in a cell. That's a far cry from caring for a patient. As a 19-year-old kid just out of his freshman year of college, this was all pretty freaky.

So when I arrived at work that night, my nurse friend Sylvia, whom I had worked for all summer, asked me if I wanted to do something a little different. When I asked her what she had in mind, she said they needed a male to "babysit" this guy with the DTs. Not knowing what that was, I asked her about it.

After the worrisome explanation, she said, "Don't worry. They'll have him tied down. You just need to make sure he doesn't get violent or get away."

This felt like a setup. After all, I wasn't a black belt, nor did I have a clue about how to keep this guy from going on a killing spree. I thought that was what police and security did. She assured me that he would be in some type of straightjacket, and I would just need to sit there. I was reluctant, but adventurous. So I thought it might be a good part of the "hospital summer experience," and I figured it had to be better than bedpans and enemas.

I walked into the room and saw a guy about 25 years old and fairly short in stature, but stocky. When I initially arrived, he was sawing logs and had no idea I was there. Easy enough, I thought. So I sat down in the dark and tried my best to stay awake. Within minutes he arose and made sure I wouldn't fall asleep--not what I wanted. Would he get violent? I had a lot of awful preconceived notions of what could happen.

When he did sit up, I could see his predicament. His arms were inside something like a straightjacket, and there were long tails coming off the back of his shirt that were tied to the bed. It was reassuring, because it looked as if he couldn't go anywhere. He started talking. At first I couldn't make out a word. It was a long ramble about people and places I had never heard of. But the priceless part about it was that every time he would try to get out of bed, he thought someone named RJ was sitting on his shirttails. I heard it over and over again: "RJ, would you get off my shirt? RJ, let me up. I'm not kidding with you, RJ. I've had enough of this."

His tone was forceful and threatening, so I was hoping he wouldn't think I was RJ. But there was no plausible way for him to escape as far as I could tell. RJ had him gripped firmly by the tails of his straightjacket.

It was a memorable night, to say the least. I learned more about this guy than I cared to. He confessed to a litany of felonies, rambling all night long. I was the only one there to enjoy the stories, but I laughed for hours.

A couple of nights later I was back on duty at the hospital, so I decided to go see this guy. He had woken up, and he was a different person now that the toxins were mostly out of his body. He wasn't perfectly lucid, but he actually did know who he was, though he probably still didn't understand where he was or why he was there. I introduced myself and told him I was a hospital orderly. I asked him who RJ was. After he told me, I apologized about RJ sitting on his shirt all night long. I'm not sure he got it.

That summer sealed my fate. I wanted to go to medical school more than ever.

In medical school I spent the first year learning how everything in the human body works. I spent the second year learning how everything can get screwed up. In that second year, I took a class called Introduction to Medicine, which gave me my first exposure to the wards. I was taught how to get a history from a patient, and then, of course, how to examine the patient. Once all of us in the class had achieved the most basic of skills, it was time to put it into practice. So each of us was assigned a patient from whom we were to elicit a history and then perform a physical exam. Once the history and physical were completed, we were to write it all down and present it to our attending physician. As you can imagine, I was scared to death the first time. I had no idea what I was getting myself into. I wanted to do a good job, and yet I lacked any real confidence to do it. What if I missed the big picture of why the person was there? What if the patient talked so much I couldn't get in a word? Or what if this person was just flat crazy? My mind was my worst enemy.

When I finally walked into the room, there was an elderly lady in bed. As I remember, she was in her 80s and somewhat frail. She seemed to be semi-lucid; she knew where she was and why she was there. I began to get the history from this woman. She was cognitively a little slow, but nonetheless reasonably accurate. We were taught in school to first elicit the history of a person's main illness. Then we were to get all the other facts about past surgeries or other medical problems the patient currently had. On that day, I got to the part on my to-do list that asked about medicines she took. She said she took just one but didn't remember the name of it. I asked her what it was for, and she had no clue.

Believe it or not, I've since learned that there are countless people who take medicines and have no idea why they're on them. She was one of them. So I asked her if she had the medicine in the room. She acknowledged that she did and that she had just taken her pill a little before I got there. She said she took half a tablet a day. She pointed me to it, but all I could see was a package of Efferdent for her dentures. Then I noticed that next to the box was half of an Efferdent tablet.

No way, I thought. No one would take an Efferdent tablet and not know it, would they? So I grabbed the box and her pill and took it over to her bed. I asked her if this was her medicine. I figured there had to be some other medicine somewhere.

"Sonny, that's the medicine I took," she said.

Just then, intense laughter began to well up inside of me. You know those situations where you know you can't and shouldn't laugh, but that just makes the giggles all the more difficult to control? Well, like busting a gut in the middle of church, this was one of those times. This woman was swallowing her denture cleaner.

My laughter was about to erupt. Remember, this was my first trip to a patient's room and my first history. I didn't want to be disrespectful, and I didn't want to fail the class. If laughing in a patient's face got back to my attending physician, I figured it would be curtains for me. But it was similar to one of those moments when you know you're about to lose your lunch, and if you don't run, it could get messy. So I ran out of the room. When I hit the hall, the laughter just poured out. Fortunately, she couldn't hear me, and there weren't many people in the hall. To my knowledge, no one noticed me.

Every time I started to walk back in the room I would laugh hysterically. But after around 10 minutes of total loss of control, I regained my composure and walked back in the room, telling her I was sorry but I had received a page. I don't think she ever really knew what happened. Since then I've improved on controlling my mind from erupting with laughter in the context of medicine. I guess after you've heard a million funny words, human noises and patient comments, you start to become immune.

Several other episodes from training still stick with me. One of the scenarios involved a drug user. Just about everyone in the health-care profession deals with narcotic seekers and abusers to some extent. You could write an entire book about the ways in which they try to trick you into providing drugs. But the most bold drug seeker I encountered happened to be one of my first. Since I was still in school, I think this particular seeker figured I was just some dumb student who didn't know any better. And in many ways he was right. I was very naive.

It all started with a patient who had an ulcer on his private parts called a chancre. A chancre is the first manifestation of syphilis. After I had seen him and we had done all the tests needed, we made the diagnosis. I went back into the room with my resident to tell the poor guy what he had caught. I then got a page, so I walked outside to get to the phone. A family member stepped out of the room and followed me. He obviously wanted to talk to me in private.

"Hey doc, can I talk with you a minute?" he asked.

I was thinking he must have had a private question about the diagnosis we had just made. Maybe he wanted to know how the guy caught it. I was willing to help, but remember, I was just a dumb medical student. My bank of knowledge was low compared with my resident's.

"Can I talk to you about my problem?" he asked. I nodded.

"I've been having these bad headaches lately," he said. "They bother me all the time. I've tried everything and nothing seems to help me. I need some help. And I need to know if you can help me. I've been asking around, and this dude told me that some medicine called heroin would help. Do you think you can write me a prescription for heroin? He said it helps him."

This guy was obviously not a Nobel Prize winner. In fact, he was a few tacos short of a fiesta platter. Everything in me wanted to start one of my previously described laughing spells. But this time, when it started to well up in me, I gained control, because, well, it's guys like these who carry very big guns. So I concluded that laughing might not be the best idea.

I told him that prescribing heroin would not happen. It wasn't going to happen then, and it wouldn't happen the next day if he asked again. He probably didn't realize I didn't even have my medical license, and even if I did have it, a common physician obviously cannot write a prescription for heroin. I advised him to take Tylenol or Advil, but I have to tip my hat to his boldness. It was impressive, but stupid.

Other times, as I've noted, I fail to maintain my composure in light of an overwhelmingly funny situation. There was another incident in which I had performed a hysterectomy on an elderly woman and was coming around the next day to see her. She was pushing 90 but very with it. As I make daily rounds with all my patients I ask a lot of pertinent questions. As I went through my questions, she said her previous night had been reasonable. Then

I did a cursory exam and looked at her incision. I had just cut her open the day before, so I wanted to check the area where I had cut her open.

When I pulled the covers back, I had to look for a few seconds longer than usual just to recognize what it was I was looking at, and to confirm I wasn't seeing things. There, precariously placed on her incision, were her dentures. They looked as if they were biting her on the incision.

Instinctively, I blurted out, "Ma'am, I think something is biting you."

Before I realized what I had said, the words were out. But hey, it was true. Her own dentures were biting her incision. I just laughed, and she turned red. She said she had been looking for them for a while. I guess she was looking in all the wrong places. It must have been the morphine playing tricks on her mind. You begin to see why morphine is a controlled substance. It can do crazy things to the brain.

As a fellow in the final two years of training, my responsibilities changed. Not only was I responsible for the patients, but I also was entrusted with educating the residents on my service. At the end of the day we would make one final set of rounds to get everyone tucked in for the evening. While we would round, I would choose a topic each week to talk about. One particular week we were discussing something called Total Parenteral Nutrition (TPN), which is an IV fluid that provides patients all the nutrition they need while they're in a non-eating state. It was getting late in the day, and I think everyone was getting tired. (Back then, the 80-hour work was not in place.) As we were winding up the night, my residents began to giggle. I didn't think much of it until the giggle turned to laughter. Finally, I asked them what was so funny.

That's when they told me to look over my shoulder. There, about five inches from my right shoulder, was the last patient of the evening, just standing there listening to my lecture. Now, the humorous part was that she had no mental faculties left, and yet she was standing there looking very intrigued. Not only that, but her gown was off her shoulders and her back end was flapping in the breeze. Fortunately, it was about 10 o'clock at night and we were the only ones left in the hallway except the nurses. I could tell the only thing that kept my residents from cutting this off earlier was that the humor kept them awake. When I realized the only reason they weren't sleeping through the lecture was an inadvertently streaking patient, I figured it was time to escort the patient to her room and call it a night.

Some might say that after a while physicians can develop a dark sense of humor, but you have to recognize the humor when it's there or you'll go insane. I'll give you an example. About 10 years ago, I got a call from the internal medicine service about a woman who was admitted for confusion and pneumonia. After she had been on the ward a few days, one of the residents decided to look her over for any signs of other problems. When the intern happened to look at her bottom area, she had an obvious skin cancer--I say obvious, because there was something the size of an orange hanging off her bottom.

So they called and asked if we could go see her. I had a chief resident on my service at that time named Howard who was in his last year of training. He was a wonderful resident, and he and I had worked together for almost four years. There was a good dynamic between us.

After we made our regular rounds, we went to see the woman with the orange. Upon opening the door, one thing became apparent. She had a positive "Q sign." The Q sign is an ominous one. If you picture a person with an open mouth and snoring, that would like the letter "O." But if there's no one home, so to speak, the tongue falls out to the side, creating something that looks like a "Q." That sign generally signals that a stroke has occurred, or perhaps some other catastrophic event. Most of the time when a person exhibits the Q sign, life is short. When I went to ask the patient a question, she couldn't be awoken. She didn't talk. She didn't even really move. She just kind of snored. (That's also a bad sign.)

So, with the nurse present, we looked at her bottom, and sure enough, there was a big cancer there. It seemed a little crazy to pursue it, but that was the request. I asked Howard if he had brought a biopsy instrument. He got a look on his face that said, "I knew I forgot something."

"Howard, how are we going to do a biopsy without a biopsy instrument?" I asked.

"Good question," he said. "I'll have one tomorrow."

It did seem like a moot point given the overall picture of the patient. She really didn't look as if there were many days left. But the next day arrived, and after we finished all of the duties on the ward and in the operating room, we went by to see our lady. This time Howard very proudly came forth with the biopsy instrument.

The nurse was present, the biopsy instrument was present, so all I needed to do was make sure Howard had consent from her power of attorney. Again he got that deer-in-headlights look.

"Do we need a consent?" he asked.

"I'm not a lawyer, Howard," I said, "but I would guess that before one can take a piece of flesh off of someone, we need some form of consent."

"You know," he said, "I'll bet you're right. I'll have it tomorrow."

It was starting to become a joke between Howard and me. Our daily routine would take us by her room at the end of all our daily tasks. I'm sure in his mind he was going down his checklist. He had a biopsy instrument. He had spoken to someone with power of attorney and had obtained the necessary consent. The only thing left to do was to get a diagnosis. I didn't know how it was going to help, but I certainly did not want to stand in the way of her care.

So we walked into the room totally equipped and she was dead. The nurse hadn't seen her in the last few minutes, but she swore the last time she had checked on her she was at least breathing.

So I turned to Howard. "I have one more thing to teach you here," I said. "It is very important that if you're going to get a biopsy, the patient has to be alive. It doesn't do us any good if we take this piece of tissue, make a diagnosis, and the patient is not alive."

"I promise the next time we'll get here before she dies," he said.

And so life goes on (except when it doesn't).

Chapter 10

Bad News Bearer

One of the most dreaded aspects of practicing medicine is delivering bad news. The worse the news is, of course, the harder it is to give. It's far easier to tell a family that their 90-year-old grandmother will not make it than it is to tell a family their 17-year-old daughter won't make it. But each case is difficult in its own way. The most disheartening part to me is that I see the utter disappointment on the affected patient's face. There are few things that make my heart dive through the floor more than seeing that expression. And when the patient begins to weep, it's that much harder. As a doctor, I feel like I've failed the patient who put her trust in me. A part of my soul makes me believe I let this person down.

Delivering bad news to a cancer patient is particularly tough. It's like destroying a person's hopes and dreams. The patient hoped to have her life preserved, and I couldn't deliver. Even though I try not to take it personally very often, it does occasionally get to me. I believe as a doctor and as a man that only God can absolutely preserve life. But there's still a part of me that wishes I could win the battle for every patient. But that is not in my power. My heart weeps, even though I know there was nothing more I could do.

It's a humbling situation. Just when you think you know how to conquer certain cancers, the cancer teaches you humility. It can bring you to your knees. It also reminds you of your own mortality. We are all going to die. It makes you realize how precious and tenuous life is. We are never more than a breath away from death.

Now before you check out of this story, there are glorious days that occur even in light of bad news. As a caregiver I'm given the opportunity to help, to encourage, to perhaps lengthen life, to know another family, to hold their hand and connect on the deepest level, and to love life while we are given that precious gift. In light of small hope, great news and celebration can occur, which reminds me of a patient I'll never forget. We'll call her Ms. Jones.

Ms. Jones was an African-American woman in her late 60s who initially went to her family doctor because of vaginal bleeding that was occurring many years into menopause. It's a disconcerting thing to happen, because obviously by that point in a woman's life, periods should be done with. But it's even more disconcerting because bleeding is an early sign of uterine cancer.

A biopsy revealed that Ms. Jones did, in fact, have uterine cancer. And not just any uterine cancer. Her cancer was called a Mixed Mullerian Sarcoma. It's "mixed" because there are two types of cells in this cancer. Those of us who deal with cancer for a living fear that name. Why? This form of uterine cancer is highly aggressive and has a reasonable probability of recurrence.

So Ms. Jones went to the operating room to have her initial surgery. At the time of surgery, we did two basic procedures. The first procedural part of the operation removed the cancer. This came in the form of a complete hysterectomy: We took out her uterus, tubes, and ovaries. The second procedural part of the operation was to determine if the cancer had spread. We obtained biopsies from different lymph nodes in the pelvic and back area, then poured saline solution in the abdomen and collected fluid to see if there were any free-floating cancer cells. All of these biopsies then went into a formula that directed us as to further types of treatments.

Ms. Jones did very well after her surgery and remained without any evidence of cancer for a short while. As she came back for follow-ups, we slowly became better and better friends. During these visits I got to know her daughters, who were her chauffeurs. One thing that became apparent quickly was her faith in God. She talked about it openly and without hesitation. This seemed to be true of her daughters, as well. As I've mentioned before, I often feel as if I become a part of many families. I get to know the kids and grandkids. I know what they're doing on Thanksgiving and Christmas. I end up sharing in their joys and disappointments. Unfortunately those disappointments tend to come from the cancer. And the hard part is, the closer I become with these families, the bad news becomes harder and harder to give. Their heartache now becomes my heartache since the patient is no longer just a name. It's one of the most difficult aspects of an oncologist's job.

With uterine cancer, one of the tests I commonly do is a chest X-ray because this cancer commonly reoccurs in the lungs. So I ordered a chest X-ray for Ms. Jones. It was positive. Before there were no spots in the lungs, but now I could see spots in multiple areas. This was a bad sign for obvious reasons, but also because this cancer doesn't tend to respond well to standard chemotherapy.

At this point, I saw the writing on the wall. I needed to have "that" discussion. Ms. Jones came in with her daughters and we discussed the pros and cons of the two main options: treatment and no treatment. She was not a candidate for surgery or radiation at this point, so treatment would have to take the form of chemotherapy, which is a medicine that goes into the veins and is designed to kill the cancer cells. Unfortunately, it commonly has side effects that can make it difficult to handle. You have to make the chemotherapy fit the cancer, but you also have to make it fit the patient. What I mean by that is, in elderly or very weak patients, you usually can't give overly aggressive chemotherapy or you'll never get them through it. They will quit mentally, and then their bodies will give up physically. Their quality of life becomes so bad that it's just not worth it to them. After all, what is the point in prolonging life if it's not worth living?

I tried to explain all this to Ms. Jones and her family in lay terms so that they could make the best decision. For Ms. Jones, it was a no-brainer; she wanted treatment. So treatment it was.

I set up her chemotherapy to be given every three to four weeks for three months. I went over her progress with her and talked about the side effects. Even though she was in her 60s and not in very good health, she tolerated the chemotherapy remarkably well and had almost no complaints. Some patients have few problems and complain constantly; others have many problems yet they never say a word. Ms. Jones was not a complainer, even though she had reason to be.

Since her cancer was now in the chest, I followed the three-month treatment with a CT scan. Generally, every few months I bring the patient in to discuss the progress (or lack thereof). When the CT scan shows a worsening of the disease, as it did in this case, it's not a fun progress report to give. But the burden falls on me, so I took a deep breath and went in to see Ms. Jones. I started with a big smile. I tried not to show my cards at first, but she knew. I think the whole family knew. It was bad news and the cancer had grown. There was no mistaking what the CT scans looked like. I tried to find something positive to say. You never want to take away all hope until there is truly nothing left to do. One of the ways I offer hope is by telling a patient truthfully that there are other options. So I explained that the cancer had grown and that we needed to look at other alternatives. Again I offered the idea of no further therapy, which would mean death in fairly short order. She listened intently to every word I said and quickly decided she had reason to live and wanted to try something else.

So I tried a different chemotherapy from the first three months. As I said to her, it was not weaker or stronger, just different. I was, as always, hopeful that we could make some progress. The problem was that I knew too much about Mixed Mullerian Sarcomas, and what I knew was not encouraging for the future. But you never know until you try, so why not? Round two, here we go.

I started the chemotherapy just as before--every three weeks for about three more months. Again, I met with Ms. Jones every month and reviewed her situation. She surprised me once again in how well she tolerated the chemotherapy. Her side effects were minimal, and her breathing remained good despite the cancer in her chest. Every month one of the "chauffeurs" was there, and every month I got to know them better, which made it harder and harder to deliver bad news. The monthly examinations with no new CT scan reports were easy. But I knew the time was drawing nigh. When the time came, the report unfortunately again showed growth.

I walked into the room for the report on round two. After the examination, I simply launched into it.

"Ms. Jones, we have done reasonably well with the chemotherapy over the last three months," I said. "You've tolerated the treatment well and had no significant side effects. But the CT scan shows that you have progression of the disease in your lungs. It is not an overwhelming growth, but it is real. I think at this point we need to reconsider where we are going."

I knew the odds were grossly stacked against us. Once someone has failed two rounds of the active chemotherapeutic drugs, any progress isn't likely. But I also knew that it was my responsibility to offer "the options," which were becoming fewer and fewer. But the patient wanted hope. So I offered another chemotherapy option and the truth about its potential success while at the same time offering hospice. Ms. Jones chose the former.

So, chapter three. Time to start all over again. This time I did something a little different. I gave her the recommendation of weekly chemotherapy. I figured if I didn't do something radically different, her time would be growing short. I went over all the side effects again. Three more months of waiting for the next CT scan. I was not hopeful.

Again Ms. Jones tolerated the treatment quite well and had no significant side effects. At the end of the three months I ordered her yet another CT scan. Given the pattern so far, I assumed the results would be bad. I didn't even want to review the outcome. But for once I was wrong. The disease in her chest had diminished in size. Was this real? I double-checked all the measurements and did the math. I reviewed the films. It was no mistake--the disease really was smaller. For once, I could deliver something other than bad news to Ms. Jones. It's the type of news that gives the patient a new lease on life, even if it's just a short one.

So walked I into the room and again tried not to show my cards. But of course they saw right through me. And that was OK, because I truly had nothing to hide.

"Ms. Jones, the areas on the CT scan show drastic improvement. We have been successful."

Words cannot describe the elation that overcame her. Huge smiles took over her otherwise stoic look. She was brimming over with excitement, and her eyes sparkled with gratitude to God.

"Praise the Lord!" she said, being the good Baptist that she was. She must have said "Amen" no less than 10 times.

As is often the case when I have a big day with a patient, it was easy to offer a prayer. So I asked her if I could pray for her. She did not hesitate to say yes. So I started into it as she held my hand. Her grip was tight. The "Amens" were flowing. Being the good Presbyterian that I am, I was not exactly used to all the Amens. But what came next really caught me off guard.

As I finished praying and started to let go of Ms. Jones's hand and walk out of the room, I felt the pressure on my hand increase.

"Doctor," she said, "we need to sing a moment."

Now, I don't sing, at least not in public (except for occasionally in the operating room). I sometimes sing in the shower, but "sing" is using the term loosely. I had certainly never, ever sang with a patient.

"Girls, come up here and let's sing praises to the Lord."

Before I knew what was happening, I was in a circle with Ms. Jones and her daughters holding hands. I could tell that something was about to happen that I would not be able to control. Ms. Jones mentioned the song we were going to sing, and it was apparent that I was expected to be a part of this experience. Never mind that a.) I didn't know the song, and b.) the walls between the exam rooms were quite thin, so everyone would hear us. But here I was, in a circle with several robust ladies starting to sway.

First Ms. Jones started carrying the main melody. The grips on my hands were increasing and the swaying progressed. Then came instruction from the mother to start in with the harmonization.

Before long they were in a pitch-perfect, three-part harmony, rhythmically going back and forth to the beat of an old Southern gospel hymn.

Being the reserved individual that I am, I was a bit uncomfortable. I'm sure my eyes were as big as donuts. If you want to talk about a fish out of water, that was me. But this was her moment of unadulterated celebration before the Lord, and I was not about to spoil it. Even though I wanted to extricate myself from the situation, I certainly didn't want to disappoint the patient, or, even more importantly, disappoint the Lord. So I went with the flow. If they swayed, I swayed. But I did draw the line at singing. I just couldn't go there.

After a while, they stopped. Ms. Jones turned to me and thanked me. For that moment, I was like family. I was quite literally in the inner circle. There are no words to describe how special that was. It's those moments that make all the bad days and bad reports melt away.

Ms. Jones lived about another year after that. We did have some other bad reports, but she took them like a champ. She knew where she was going, and she didn't fear it. She was an inspirational woman that I learned a lot from. I think of her often even now, several years after her death. She showed me that there is hope when all hope seems lost. There is celebration even when it seems impossible. There was hope even in facing death. And that was worth a lot.

Chapter 11

Southern Belle

When I receive a phone call from a referring physician about a new patient, I'm always slightly skeptical about what I'm told. Sometimes referring doctors send over patients they perceive to have an obvious cancer but in fact don't. Other times the exact opposite happens--the referring doctor thinks it's benign when it's actually malignant. And though we don't like to think of our doctors in this light, there are times the referrers simply don't want to take care of a patient. Perhaps the case would simply be too big or too difficult. Perhaps the doctor's going out of town and doesn't want to have to deal with it upon return. There are a million reasons why patients are transferred to us. Ultimately, that's fine, because my job relies solely on the referral business. If I hadn't wanted the referrals, I would have done something different in life. But after you've seen just about every kind of transfer, you can't help but be slightly skeptical from time to time about why the patient is coming to see you.

Referrals are also intriguing because you never really know what the patient will be like, physically or emotionally. Physically, the patient may be so sick when she comes through the door that any hopes of helping her may not be realistically possible. Or, someone who was described to be at death's door prior to the transfer may look like a million bucks. Emotionally, you also wonder if the patient is stable or labile. These emotional issues come to bear when the treatment is being planned. If the patient can't handle the treatment or possible surgery emotionally, then it's difficult to offer the option of intervention. Sometimes the patient doesn't want any help. Other times I can't help, and she wants me to move heaven and earth to make it all right. Then there's the patient who doesn't have all of her mental faculties and cannot realistically make an informed consent. Even worse is when that patient has no family willing to step up and help them make the right choices.

Sometimes the patient is a talker. I've had patients who would go on gabbing for hours if I didn't cut them off. I'm forced to say things that could be interpreted as rude just to get them to answer the question at hand. I think of that old line from *Dragnet*: "Just the facts, ma'am, just the facts." Other patients won't answer at all. It's not that they can't talk, they just won't. I'll beg, cajole, smile, stand on my head, tell jokes--all for a blank stare. In some cases the family answers everything and won't let the patient get in a word edgewise. The patient will start to answer the question and the family continually cuts her off. That makes it hard to know if I'm getting the whole story. Other families completely abandon a patient. When I was a student at Parkland Hospital in Dallas I would see families drop off their poor grandma or Aunt Bessie and just walk away. She would sit there in the emergency room not knowing where she was or why. It could be hours until someone realizes this person is lost and alone.

Some patients are crazy. They say things so far off base you wonder where the response came from. They tell you about the bugs on the walls. They tell you that FDR is still president. I had one patient who believed her family had kidnapped her and stolen all her belongings, and that I needed to call the police. Then she leaned in and whispered that they were out there in the waiting room right then.

Some patients come in mad at the world. Though I've never met these patients previously, I'm already the bad guy. Perhaps they had a bad experience with the previous doctor, which somehow becomes my fault. Sometimes they're mad at the potential diagnosis and they channel that anger towards me. To be fair, no one wants to visit the gynecologist, so I rate very low on the totem pole of people patients want to see.

Fortunately, 90 percent of what comes through the door is just like a friend or a next-door neighbor. One particular day, I received a referral for a woman in her 80s who had an ovarian cancer. "Margaret" embodied everything that one would imagine in a perfect Southern belle, from her lovely accent to the elegant way she carried herself. She lived locally, but her family was about two hours away in Atlanta, and her daughter would usually bring her in. I found this woman's Southern sensibilities and kind manners delightful. She always had a story to tell. Sometimes the stories got a little wordy. Actually, they were always wordy. Fortunately, her daughter was usually there to keep an eye on things. After about five minutes her daughter would have to cut off her mother, saying, "Now Mom, you know the doctor has other patients. You need to let him go back to work." The daughter's intervention became so routine that I could almost predict it every time. But it always made me smile. We had a lot of laughs that first year. I learned to always allot a little more time when she would come in.

I have a routine during office visits that I go through with all the patients. After the examination, the patient, the family and I talk about where we are and where we're going. I always end with, "Do you have any questions?" This, of course, is where Ms. Southern Belle would engage me in some conversation about the subject du jour. I'll never forget one particular conversation that took a turn she never expected.

"Dr. Puls, you are going to be so proud of me," she said. "I have learned something that I'm going to implement into my life. I just had to tell you because I think it will help with my cancer."

"What's that, ma'am?" I asked.

"I was watching *Oprah* the other day and learned that meat is not good for you."

She then launched into a story about a segment on Oprah Winfrey's show that dealt with Mad Cow disease and its implications for our society. During that show, Oprah said she would never eat meat again. Because of that comment, there was a drop in the sales of beef across the country.

For any other doctor, the story might not have been that big of a deal, but I had a special attachment to this situation. See, I was born and raised in Dallas, Texas, and my wife is a fellow Texan. In stark contrast to my big-city upbringing, she grew up on a working cattle ranch in West Texas. She was part of the fourth generation to run the ranch. And yes, they really do round up cattle on horses and do all the things you think of when it comes to ranching. Even though I grew up in the same state, it was a shock for me to see people live like that. I thought that stuff just made for good television. But indeed, it was the stuff of real life.

Anyhow, following Oprah's Mad Cow show, a group of Texas cattlemen took her to court over the situation. Oprah moved her show to Amarillo, Texas, and broadcast from there during the case. One of the men in the lawsuit was my father-in-law, a man I have deep respect for.

Margaret wrapped up her long-winded story by saying, "Can you believe those men out in Texas are involved in a lawsuit with her?"

I didn't know what to say. There was my family connection to the case, of course, but like most guys from Texas, I also just love beef. I grew up on hamburgers. Like the bumper sticker says, "The West wasn't won on salad." Beef is also a good source of protein and iron for cancer patients.

I wanted to say kind words to her, but I didn't know how to articulate them. After all, she was speaking from her heart. I didn't want to burst her bubble. She was so proud of herself for coming to this revelation, and I could tell she wanted me to be proud of her. But I had to say something.

"Now ma'am," I said, "I need to stop you for just a minute and let you know something. I am an enormous supporter of beef. If you remember, I'm from Texas and lived there for the first 27 years of my life. Most of us from Texas love beef. I also think it has its benefits when it comes to helping you during your treatments for the cancer. Plus, and I hate to say this, my father-in-law is one of the men involved in the suit against Oprah."

There was dead silence. You could have heard a pin drop. She just stared at me. I could see her wheels turning. As a proper Southern woman, she would never intentionally say a mean comment to anyone.

"Dr. Puls, you're kidding me, aren't you?" she asked.

I knew she was hoping this would all go away. Still, she actually had not offended me in the least. In fact, I was laughing on the inside.

"I hate to say this, ma'am, but that is the honest truth."

She contemplated her next words carefully. "Well you know what, I will eat beef from now on. I love beef and I'm not really sure I believed that information anyway."

Finally, the tension was gone and we both had a good gut laugh. I think the whole incident ended up bonding us more closely.

On following visits, she asked me more and more about Texas. She wanted to know about its history and its flavor. At one point she asked me if I knew of any good books on the state. Coincidentally, I had just finished a historical fiction novel (Not Between Brothers) set during the early years of Texas. It was a story filled with cowboys and Indians and all the interactions between the two groups. The book focused on two half-brothers, one cowboy and one Indian, who had lives paralleling each other. Woven into this account were all of the events that led to the formation of Texas into statehood. But it was a tough book with respect to all the killing and other gruesome events. You needed to have a strong stomach. But it was a book that moved me deeply.

So when she asked me about a good book, I just blurted out the title. Then I thought to myself, "What are you doing? She could never read a book like this. There's way too much cruelty, violence, etc. This is not the book that genteel Southern women would read." But I had already let it slip. This, of course, led to another five-minute conversation. I tried to backtrack and explain that this book might be a bit violent for her.

"I've read a lot of books in my life and I'm sure I'll do fine with this one," she said gracefully. She had made up her mind and

I wasn't going to change it. I brought her the book and hoped it wasn't an awful mistake.

Not a week had past before the book was back on my desk with an eloquent thank-you note. Even her handwriting was beautiful.

There was one other particularly memorable office visit. After a regular checkup, her eyes began to well up with tears. I had no clue what was wrong. I racked my brain to figure out if I had said something wrong. There was definitely an elephant in the room, and it needed to be acknowledged.

"Dr. Puls, I need to talk to you about something," she said with some hesitation. She followed with a long explanation, as was her custom. I'll give you the abridged version.

"I have disappointed you, and I am very sorry about it," she said. "You know when I was here last month? Well, something happened after that and I need to tell you about it. Since you are from Texas and love beef, I know you will be particularly disappointed".

"When I left here after an appointment, I was driving home alone. I had just gotten off the interstate and I was on the service road. While I was driving, I got a little confused. In my confusion, I drove off the road and went through a barbed-wire fence. On the other side of that barbed-wire fence was a large field and as I was driving through it, I couldn't seem to come to a stop."

As a side note, I had given her permission to drive, but only in the *medical* sense. That is, if she was OK to drive *before* treatment, she should be OK now. Some patients, however, equate my approval with that of a highway patrolman. I'm by no means someone who can or should judge a patient's driving *skills*, but sometimes they take it as such. It's sometimes scary to think of some of my patients out on the road. This woman was one of them, and her story proved my fear to be well-founded.

She continued with her story, now sniffling. "Because I couldn't stop, I did something awful. I know you're going to be very sad. In that field were a bunch of cows. I finally came to a stop when I ran into a cow. And the worst part is, I killed that cow."

She was silent for a moment, before ending with, "I know how much you like cows and I am sure that I have let you down."

I was reminded of times when my small children would cry about something that seemed so minute, but you couldn't laugh. To them, the event was huge. And this event was a big deal to her. I didn't dare chuckle for fear that I would offend this dear soul. I just needed to clarify the offense.

"Are you telling me that you think I will be mad at you since you killed a cow?" I asked.

"Yes," she answered.

"And the reason you think I'll be mad is because I'm from Texas and my wife comes from a ranch?"

"Yes."

I had to quickly think of something awful or disgusting to avoid busting a gut. I was absolutely on the brink of falling down on the ground in uncontrollable laughter. I don't have any personal feelings for cows. After all, I have them for dinner and lunch all the time. But I could tell that her confession was cathartic for her. It was as if a giant burden had been lifted from her. She believed that she had let me down and her conscience could now be at ease.

"You know, ma'am, you don't need to worry about that," I said. "I am truly not mad. I just want to know that you're OK and didn't get hurt."

She was very relieved. Now that she had spilled her guts, she felt redeemed and forgiven. She and I were still friends. We could still have our regular meetings and long conversations as her oncologist and social outlet.

Of all my patients, this woman was most endowed with the gift of gab. And if it hadn't been for her daughter, my visits with her would have been day-long experiences. But I admit, spending a day with her would have been a true delight.

Chapter 12

Back to life

Watching someone die from any cancer is difficult, but watching a person die from cervical cancer can be particularly hard to take because it's known for its ability to cause overwhelming amounts of pain. That pain results from cancer's tendency to grow through structures that may have a lot of nerve supply, such as bone and muscle. Cancer has no respect for normal structures. It sometimes loves to literally eat right through them, down into the nerves. The pain can be so intense that patients may simply wish for death. Often I'm relegated to giving so much narcotic that the patient cannot even see straight.

This is the story of a patient who was in just that situation. "Jessie" was a young German woman of 33. Before she was my patient, Jessie had a few abnormal pap smears, one of which led to a cone biopsy. This is a procedure where an ice-cream-cone-shaped area is taken out of the middle of the cervix. It's done to remove all of the abnormal areas from the cervix. Jessie's report showed that all of the abnormal areas of her cervix were indeed removed, and her pap smears reverted to normal.

Within a few years of this incident she conceived her first pregnancy. Her pap was normal early in the pregnancy. There were absolutely no warning signs. The baby was growing and moving around. Life was good. As she began to enter the final weeks of the pregnancy, her obstetrician checked her cervix to see if any dilation had begun. As Jessie later told me, it seemed odd that the exam was taking so long. There was a puzzled look on the doctor's face.

"We need to do an ultrasound," the obstetrician said, "because I think the baby is breech. There is a very unusual feel to the cervix that makes me believe the presentation of the baby is not normal."

But the ultrasound revealed there was no breech presentation. At this point the obstetrician wisely decided to take a look at her cervix. Then came the shocker. It was cancer. In the blink of an eye, Jessie had gone from finding out how soon her baby would arrive to hearing a potential death sentence. I can only imagine how she must have reacted: silence, tears, disbelief... Maybe it's all a dream and will just go away. But it was real; the biopsy confirmed it. Not only was there a malignancy on her cervix, but it was big enough that she would not be able to deliver vaginally. If a cancer of the cervix gets to be a certain size, you cannot allow the baby to come through the birth canal because it will cause the cervix to bleed. Once a cancer like that begins to bleed, there is no real way to stop the bleeding short of taking out the uterus. It is a general rule that one cannot sew into a cancer because the cancerous tissue will not hold the suture.

Her obstetrician had a wonderful reputation and I had known her for many years. She called me, devastated. I asked her to let me see the patient, and we would create a plan. That was the first time I met Jessie. She was a strikingly beautiful young woman--moderate height, slim, with blond hair. But I could also see the fear in her eyes. The look on her face was more than I could bear, given the apprehension over the impending delivery of her firstborn and the news of the cancer. It was hard to know how to comfort her. Her supportive mother came with her, as was her habit. (If there was a husband, he was not to be found.) Though Jessie had a minimal but ever-present German accent, her mother's accent was thick, and she asked most of the questions because there was some denial for Jessie.

After some time talking, we got the examination underway. From the moment I eyed the cervix, it became grossly evident we had an uphill battle. The size of the cervical cancer was about five centimeters, a reasonably big cancer of the cervix. Biopsies were obtained and a plan was set into motion: schedule a cesarean section and then obtain a number of lymph nodes to see if the cancer had spread. I could tell she was numb. She didn't have a lot of questions. I think she simply didn't know what to ask. What does one say to this kind of news? Her life had radically changed in a moment and the hopes of her future were in question. But at this point I didn't have all the information I needed. There were X-ray studies to do, lymph nodes to get, and a baby to be born. Her eyes welled up, and I comforted her as much as I could with the little information I had.

The treatment for cervical cancer is generally either a surgical treatment or radiation and chemotherapy. There are many more nuances to its treatment, but we'll keep it simple. Surgery is for people that are reasonably healthy and have the cancer confined to the cervix alone. The cervix also needs to be fairly small, with an average cutoff of four centimeters across or less. Radiation and chemotherapy are given for cervices that are bigger than four centimeters and/or the cancer is no longer confined to the cervix. Given those basic guidelines, Jessie ultimately needed to have radiation and chemotherapy.

On the day of the cesarean section she delivered a healthy, beautiful baby girl. After they closed up the uterus, we began our work as oncologists. A number of lymph nodes were taken out. Unfortunately some of these were positive for cancer. With each cancerous node that came out, I began to realize her chances of survival were diminishing. At the end of the operation, we closed her up and let her heal. About three to four weeks after the surgery, she came back to start our treatment planning. We began radiation and chemotherapy without delay. The radiation is given in two parts: The first part is given from the outside and delivered into the pelvic area for about five weeks, and the second part is given as an internal implant placed directly into her cervix. Fortunately, all of this went very well. There were no delays and no treatment problems.

After all the treatments had been given and Jessie was recovering from the radiation and chemotherapy, a CT scan gave us what we were hoping for--a good report. There was no evidence of disease. Her cervix looked to be without cancer and she was slowly getting her strength back. The pap smears had reverted to normal. The radiation had few physical effects on Jessie's body, and the chemotherapy had not taken out her hair. She was, for the most part, unscathed from the therapy. Her baby was growing and beginning to smile. For the first time, I remember seeing a look of hope in Jessie's eyes.

At this point in cancer management, we commonly go into an observation mode. With cancer of the cervix, we manage the patient with pap smears, CT scans, and examinations. In patients that have a recurrence from their cancer, the recurrence will often be in an area near where the original cancer began. Thus a lot of pelvic exams are done, and that's what I did for Jessie. She had multiple follow-up appointments over the next 12 months.

Everything was going as I had hoped until she showed up one day in pain. New onset of chronic and progressing pain is an ominous sign in cervical cancer. As I mentioned, cervical cancer loves to grow into and through adjacent structures. When any cancer has that pattern, it tends to cause unrelenting pain that's not associated with any particular activity. It's just always there. Jessie's pain led to an exam, which led to a CT scan, which led to a biopsy. The biopsy revealed the very thing she and I feared the most. She had a recurrence. The skin over the area in the pelvis where the cancer was residing began to dimple. The recurrence also caused swelling in her leg. Life had changed for everybody. It changed for Jessie, her daughter, and her family. With cervical cancer, some recurrences allow for a surgical option. This was not one of those. Other recurrences allow for more radiation. This was not one of those, either. Then there are recurrences that only allow for chemotherapy but will inevitably lead to death. This was that kind. It wasn't a matter of *whether* she would die, it was a question of *when* she would die. It's a particularly tough fact to face when the patient is a woman in her early 30s who has a 1-year-old at home. Who would raise the child? It disheartened me to see her daughter and know that she would never really know her mother.

But I couldn't let my mind go there. I understood the reality of it, but I couldn't dwell on it every time I saw her. It would be difficult to dwell on.

After the biopsy, Jessie and I sat down and discussed the chemotherapy options. She consented to go on treatment. But with that treatment, I had to be honest with her and let her know that ultimately this cancer would take her life. The goal was only to prolong her time here on earth. We were not after a cure. I hate having those types of discussions. The most difficult talk of all is when a patient is young and has children who depend so heavily on her. To see that little toddler in the same room and know she will grow up without a mother is horrible. The day was difficult for both of us. But Jessie didn't cry as much as I thought she would. Her denial, and particularly her mother's denial, was overwhelming. I don't think her mother ever really got it, and she fed that same thought to her daughter.

The chemotherapy occurred about every three weeks and the goal was to potentially shrink the cancer. I did not give Jessie unrealistic hopes that she would likely go into remission. She understood that this was purely a palliative treatment plan to buy time. At least she said she understood that. Occasionally a patient has a remission from this cancer with chemotherapy. One patient of mine has gone into remission three times with chemotherapy for her recurrent cervical cancer. So it does happen, just not very often. Certainly it doesn't occur as often as I'd like.

Jessie's cancer did quit growing for around four to six months. During those days, her quality of life wasn't perfect, but it was good enough for her to care for her little girl. In a sense, they were sweet days. She was alive, and her daughter was healthy and happy, still oblivious to the impending nightmare soon to come. Eventually the pain progressed, and the use of narcotics began, which is a blessing and a curse. The blessing is that the pain is not as bad, so the quality of life is better in that respect. The curse is that her sensorium, or the sense of the world around her, changed. She could not think as clearly as she wanted to, so the times with her family were not necessarily as good as she had hoped. There was no easy answer. Medicine is not always black and white.

The recurrence was located deep in the pelvis in an area that had already had radiation. When the chemotherapy started to fail, the writing was on the wall. I knew her life would not last much longer. The cancer was invading, attacking the blood vessels that supplied her left leg. It began to swell and become discolored. The pain became excruciating, and the use of narcotics went up so much that she became somewhat confused. But she needed control of the pain. Eventually she had to be hospitalized in order for me to help her. The sweet person I had come to know became an all together different person. She was still sweet, but she just wasn't totally there as the narcotics robbed her of some of her faculties. I tried various pain-control options, but finally placed her on IV narcotics in the hospital. Once she was in-house, I realized how quickly she was losing her life. Every day she looked weaker and more emaciated. The color had left her skin. The beautiful and vivacious young woman I had come to know only a year and a half ago was deteriorating before my eyes. The skin over the cancer began to erode and a large ulcer formed. Her leg had swelled to at least twice its normal size. She had gone from a healthy 130 pounds to well less than 100. With the exception of her swollen leg, she was indeed skin over bone. Her eyes began to sink into her head. The smile disappeared. That funny little accent of hers was barely audible. I don't entirely know where Jessie went, but from my vantage point, she was gone. I really could not recognize her. It was just a matter of time.

It was difficult enough to be a bystander and observe this progression, but when her daughter came by every day to see Mommy, it was that much harder. Jessie's mother cared for the child and would transport her each day. The mother would always ask me for an update, but when I would tell her my thoughts, she didn't hear them. Denial is so powerful and hard to understand until you see it lived out. After our conversation, she'd ask when Jessie would be well and able to come home. Inside I would just shake my head. You can't force the truth on someone who doesn't want to hear it. You just consistently present it and hope that at some point it sinks in.

In a hospital setting, when a patient such as Jessie is dying and it's clear that life cannot realistically be prolonged, we obtain what is referred to as a "code status." This code status lets everyone in the hospital know whether or not an individual should be resuscitated in an emergency. You see, in hospitals there are teams on call 24/7 for patients who quit breathing because the doctor who admitted the patient may be elsewhere when that happens. Also, doctors who do not run code events on a regular basis shouldn't run them anyway. And frankly, they don't want to. But it is the primary physician's responsibility to inform everyone as to the code status. This way if a patient stops breathing and the patient does not want to be resuscitated, then the code team is not called. What you don't want is for the team to be called, revive the patient and put him/her on a breathing machine when the patient was at peace about dying. It's important to address this issue with patients and their families. No one wants to discuss this subject, but it's a necessary difficulty. Having said that, I attempted to talk about this with Jessie and her family, but when patients are in denial, it's even harder to talk about. As you may guess, the family wanted everything to be done to revive Jessie. So she was considered a "full code." It wasn't what I wanted, but it was reality. I could appreciate where they were coming from being that she was a young mother, but I believed it was the wrong choice.

When the fateful day arrived, Jessie was merely a shadow of her former self. The ulcer in her groin area had grown many times over as the cancer literally began to eat away at her body. For several weeks it had oozed minimal blood. I simply had the dressing changed on a daily basis to prevent it from getting on her legs. She had lost a lot of her color and the pain medications had taken much of her cognitive skills. Inside I said, "I wish you could see the reality of the situation. A code is just a breath away." But she and the family didn't want to discuss it.

After finishing rounds that day, I went to my office to see other patients. Then the page came. In a distressed voice, the nurse came on and said that a code had been called. A nurse had found Jessie unconscious and bleeding profusely from her groin area. I grabbed my white jacket and ran as quickly as I could to get there. Jessie's room was about six to seven minutes from my office. Jessie's mother and daughter were just arriving outside in the hallway for their daily visit. Inside was a room full of people doing chest compressions and giving her oxygen. There was an overwhelming amount of blood on the bed and the floor. She was bleeding to death in her bed. Every push on her heart just forced blood out of the groin. To stop the bleeding required not pushing on her heart, but to not push on her heart meant she was dead. Unfortunately, the scenario was obvious from the moment I walked into the room. It was a no-win situation. She was within minutes of dying and there was not one thing that I could realistically do. There was no stopping this bleeding.

I did not want to run the code because it was not something that I ever did anymore in my career. But I did have much to offer. I walked up to the doctor running the code and explained the big picture. The wound could not be stopped and the patient had a terminal cancer that was inevitably taking her life. I explained that I had wanted a code status from the family, but that was not their wishes. I honored that of course. But after thirty minutes of doing everything we could honestly do, I asked the team to stop and thanked them for everything they had done. I had done the very thing the patient and the family had wanted, but it was time to let go. And so the beautiful young mother of one died. The Lord alone could have stopped it. I had done what I was trained to do and it had failed. Now I had to tell the family.

It's one thing to tell a family their 80-year-old grandmother has died, but it is a wholly different thing to tell the family a 30-year-old mother has died, and her child is now without a mom. I was truly dreading it. I knew the mother and the little girl were just down the hall. I wanted to crawl into a hole and hope the whole episode would just disappear. But I knew it wouldn't.

I walked out into the hall. The nurses had anticipated the worst and asked if they could hold the little girl. At this age she would not understand the concept of Mommy never coming back. I walked down the hall to where Jessie's mother was, but she ran out of the room towards me when she saw I was coming. I had hoped that she would have stayed in the room to keep this moment as private as possible. We could have sat perhaps on a bed or in the chairs, but instead we were smack dab in the middle of the hall. There were so many people there from the code team and passersby who were trying to figure out what all the commotion was about. It was about the worst place imaginable for this condolence to be delivered. I started by telling her that her daughter had begun to bleed from the groin and that no matter how much pressure was placed on it, the bleeding couldn't be stopped. Then I explained that her heart had stopped and everything possible was done to revive it. And then with a great big gulp, I said she had died and was no more. She was gone. That beautiful soul that I had only known for about 18 months was gone and I couldn't bring her back. I would never see her smile again and never watch the interaction between Jessie and her child.

Jessie's mother then grabbed me on the lapels and started shaking me in the middle of the hall. "She's not dead, do you hear me? She's not dead!"

With each repetition, she kept shaking me. Desperately, she then said, "Go in and wake her up and bring her back. Don't you see, she's not dead!"

The shaking of my lapel went on for at least one to two minutes. I couldn't stop her and I didn't try. She wasn't really mad at me. She was mad at life. She was mad at the world and I let her vent. She was simply overwhelmed with unexplainable sadness--a sadness and pain that were so deep I'm sure my words fall short of expressing it. I'm not one to get choked up, but this one got to me. I was hearing the mother and seeing the little girl go by in the arms of the nurses. It was indeed too much for anyone to bear.

Finally, at the mother's request, I went back in Jessie's room and came out with the same result. This time I just held her and gave her my shoulder to cry on. It was all I had to offer.

But where does one go from here? Doctors have feelings, and in this case my senses were stretched too far. I wanted to crawl in that proverbial hole and hide. But for the moment I had to clean up the details on the chart. I had to bring some semblance of order back to the room. I had to set plans into motion for the mortuary to come. I had to help Jessie's mother think. But none of that lessened my desire to run away and hide, and for that short while there was no hole. I had to remain strong. It was important for the entire team to show strength if, for nothing else, to help the family.

When it was all over, though, I did crawl in that hole. I went back to the office and did some soul searching. Is this really what I should do for a living? When most people have a bad day, maybe they are overworked and get home late. But in medicine, the bad day can be emotionally unbearable, for below the tough exterior lies the heart of a human. And we all share that most sacred of emotions that cherishes life. Those lives we cherish can be young or old, funny or serious, big or little, but they are the ones we care for. It's easy to get attached to certain patients and it's hard to die with them. It was difficult indeed to watch Jessie pass from this life to the next. It was tough to watch that little girl running in the hall not having a clue about what had just transpired. But then you grab your regular coat, jump in the car and go home, and then come back the next day and start all over again. And hope that you can walk away from all the memories that seem to stay with you forever.

Chapter 13

Taken Down on the Elevator

Fame is a difficult thing to understand. As an ordinary man, I see the glitz and notoriety in Hollywood and find myself partially seduced by it. But the fame of these stars and starlets makes their freedom to come and go as they please tenuous at best. They have to be careful where they go, when they go there and who they go there with. How often do you and your friends enjoy a meal at a restaurant? When you go out to eat, you enjoy the fellowship and conversation without worrying about being interrupted.

When practicing medicine in a community, more and more people begin to know who you are, either by reputation or by simply recognizing your face. You become a celebrity, in one sense, in your own community. The longer you're involved in medicine in a community, the longer you have the chance to touch the lives around you. You get the sense that wherever you go, people may recognize you. That can be good and bad. This is the story of a time when it was particularly bad.

This incident is truly one of the most unforgettable situations I have ever been involved with in my career. It makes me realize that in many ways I live in a glass house. You see, lots of people have the advantage of knowing who I am, but I don't always know who they are. (Perhaps my memory isn't as good as it used to be.) I run in to people in restaurants, malls, and the like, and they ask me about a relative I took care of sometime in the past. I'm sure I look like a deer in headlights as I search my brain to recall the person, usually to no avail. I tend to say, "Yes, I remember," but in reality I have no clue. I know I shouldn't lie about it, but I don't want the family to think I didn't care about their loved one. Which, of course, isn't the case. It's just a lack of memory. Out of the office environment it can be difficult to remember who the patient and family members are. If I had a chart and a picture in front of me, the memories would all come flooding back. But I don't generally carry those around with me in the mall.

On to the event. It had been a long morning. I was leaving rounds with my residents, and about a million thoughts were going through my mind. I asked myself, as I do every day, about every lab, every X-ray, etc., making sure I had crossed every "t" and dotted every "i." I'm sure that on many days I have walked by friends or colleagues and never even seen them. I get in that zone where I don't think about anything else.

My office is in a cancer center attached to the hospital, equivalent to the third floor of the hospital. Although I do walk up stairs sometimes, I usually limit that to about one to two flights. On the way down, though, I will often go all six floors. It's not just for my health; at the rate the elevators travel, it's far easier to walk than to ride.

After noticing there was no one waiting by the elevators, I decided on this day to take the elevator down rather than walk. I generally would have shied away from the elevator, because in that setting I have been asked just about every question imaginable. When you're alone with people in that enclosed space, they'll ask your advice about their sore throat, broken bones, etc. I even had one guy ask me about his girlfriend's diarrhea. It's one thing when I'm in the privacy of my office; after all, that's why I'm there. But on a public elevator? I guess they think free medical care just walked into their presence. Sometimes I say they should go see their doctor, and then they act put out that I won't help them. So you get the idea as to why I am generally reluctant to board the elevator by myself.

As I waited for the excruciatingly slow elevator, a boy walked up to the door near me. He had a certain look to him that suggested he was being raised in a rural setting. When the door opened this young man, about 10 years old, got halfway across the threshold of the elevator door and stopped. I was taken a little off guard but put no more real thought into it. So I was in the elevator, he was at the door, and the elevator could not start while he was there. At first I thought he was just some kid trying to have a little fun with a doctor. Since I wasn't in a big hurry, I didn't pursue it. But after about 30 seconds I decided to ask him politely why he was leaving the door open. With a significant dialectic tone, he said he was waiting for his family. About another 15 seconds went by and finally "the family" made its way onto the elevator. In this group were two women, one around 40 and the other about 55, and a teenager who was the boy's cousin.

It all seemed harmless enough at this point. I was only going to ride down three floors with four other people. How bad can things get in maybe one minute?

But as soon as the doors began to slowly close, the woman in her mid 50s approached me with complete disregard for my personal space and physical comfort zone. She was no more than five inches from my chest. She was a short woman; I'm about 6'1," and her head came right to my chest. On my white jacket, my name and occupation are written at about her eye level.

I began to get really uncomfortable, and lots of questions were racing through my head. Who is this woman? Why is she so close? Do I have a goober on my tie? The woman stared at the monogram on my chest, looked up at me with a somewhat inquisitive look, and asked, "Whatta you do fer a livin'?"

It wasn't what I had expected to come out of her mouth, but then again, I didn't know what to expect. By this point, the doors had closed and it was just the five of us. All eyes turned toward me. "Not to worry," I thought. Even if I tell her, she probably won't understand what it is I do for living. I have *good friends* who can't seem to figure out what I do. The answer would probably squelch the conversation. So I said, "I'm a gynecologic oncologist." I figured that would be the end of our dialogue. But it was not to be.

"I knew that's what you was," she said.

Apparently I didn't have this situation figured out. While standing there semi-perplexed, I began to feel the elevator slow down. Although I didn't know what floor it was stopping at, I knew it couldn't possibly be my floor. So I was potentially in this situation for another 45 seconds. How many weird things could happen to me in a mere 45 seconds?

While still very much inside of my physical comfort zone, the woman said, "You took care of my daughter and she had cervical cancer."

In that statement it was clear she really did know what a gynecologic oncologist was, and that I was indeed that someone. I went from a position of feeling somewhat in control of the conversation to feeling utterly puzzled. Hopefully this was just a simple situation where we would discuss her daughter's well being.

Then a confounding element came to bear. We came to a stop and the doors opened, revealing a sight that brought me chest pain. No less than eight people were standing at the door waiting to come in. This severely handicapped me in several ways. For one, I had to move to the back of the elevator, pinned against the wall. There was no escape. Of course she followed me to the back of the elevator. I was a caged animal. My heart rate was going up. Also, when eight people hop on, the doors never seem to close. It seemed an eternity. Plus, as our new guests entered, the woman took the conversation to a new level, both in volume and in content.

"You told me my daughter would beat the cancer."

I can't remember every conversation I ever had with every patient, of course, and I certainly didn't remember the conversation I had with this woman's daughter. For that matter, I couldn't remember *anything* I said to to this woman or her daughter. But I do have a rule in medicine that I have always lived by: I never, ever guarantee a patient will beat the cancer. Only the good Lord can make that promise. No matter how good an oncologist thinks the odds are, cancer can always throw a curve ball. I never promise anyone a perfect outcome. That doesn't mean I take away all hope.

I can truthfully let patients know when their chances are excellent without promising a cure.

"You told me my daughter would live, and she died. She died."

She went on to repeat "she died" at least half a dozen times, and each time she added a few decibels and more harshness.

How do I get out of this one? The lioness had pounced, and I was feeling the grasp of her claws taking me down. It was hot, stuffy and uncomfortable. Then it got worse. She pointed emphatically at the boy who had held open the door and told me he was the patient's son. Her words and her tone implied that since I couldn't cure her daughter's cancer, this boy should be my fiscal responsibility. My responsibility! Now, I am not insensitive, and I could certainly feel her pain, knowing that seeing me brought it all back. For her, I was a nightmare. I am the one who supposedly made the promise and didn't keep it. But in that moment, I was not sympathetic in the least. I was suffocating and looking for a way out. The last thing on my mind was her pain. But there was no apparent way out. It was truly one of those moments when you just want to crawl in a hole and disappear. I had no way to defend my position with someone this emotional and irrational.

Then a miracle happened. The elevator stopped. The door opened, and fresh air was pushing through. I could breathe again. I took on the mindset of a fullback, looking for a way out through the holes and spaces between people. I calculated angles and directions and plotted the course. I didn't know what floor it was, and frankly I didn't care. I was going for it, because it was now or never, and now seemed like the better alternative. So, like a bull in a china store, I barged my way through the elevator without any rebuttal to the woman. I could feel my heart racing. It was almost as if the woman was chasing right behind me. Realistically, I knew she wasn't, but your mind can play weird tricks on you at moments like these.

Finally, I was free. My sentence in that prison was only about one, maybe two minutes, but it had seemed like life without parole.

Occasionally, when I think back on this story, a cold sweat breaks out. The whole incident has actually had some lasting effects on my behavior. I rarely ride the public elevator at the hospital anymore. I've also adjusted how I walk down the corridors of the hospital, and even how I make eye contact when I'm out in public. Somehow my mind won't let me forget that day. I don't want to be seen, I don't want to be recognized, and I don't even want to start a conversation with someone lest I get stuck and have no way to free myself. I guess that's life in a glass house.

Chapter 14

Spoiled Brat

I often come across articles that talk about how broken our health-care system is. These articles rarely point out the system's strengths. But if it's all so bad, why do so many people come to the U.S. for surgery when they could undergo the same procedure at home?

The answer became quite clear when I went to a Third World country to operate. Even though the health-care system is certainly not perfect here in the states (for a multitude of reasons), it is, in many ways, the envy of the world. Operating in a Third World nation showed me what a spoiled brat I am.

For a long time I had considered doing something out of my occupational and cultural comfort zone. My good friend Richard—the pastor from a previous church I attended—and I had often considered doing a mission trip to a Third World country. We believed we ought to give back to the world a small piece of what it had given us. Plus, we thought it would be good to try something that would truly make us stretch, and we figured giving up creature comforts like the Golf Channel and the golf course would help.

Richard and I discussed it off and on over the years but didn't pursue it whole-heartedly. Every now and then he would forward me an e-mailed proposal, but it always involved working at some clinic doing basic family medicine. I wasn't trained to do that, and I knew I would be horrible at it. It had been so long since I had taken care of things like pneumonia and blood-pressure issues that I probably would have been dangerous in that setting. I didn't want to go somewhere where I could get arrested for botching something and then get thrown in prison and never see the light of day again. I wanted to be able to truly help people in what I was trained to do. And, to be honest, I was also just curious to see what medicine was like in another part of the world.

Then one day Richard came to me excitedly.

"You won't believe this," he said, "but an organization has come forward with a request for someone to do gynecologic oncology. And the request is from a Third World country."

It seemed too weird to be true. After all, how many communities are in dire need of a gynecologic oncologist? This had to be the trip I was destined for. Richard pushed me on it, saying, "Sign up or quit asking." So we signed up. We were going to the Ukraine. I figured Ukraine was a reasonably advanced country, having come out of the Soviet Union, but when I got there, I realized it was anything but that.

We researched our task and realized we'd need other players on the team and some supplies. It was a daunting task. We would need someone to put the patients to sleep and a circulating nurse to complete the group. Then we had to talk those folks into coming with us.

Next, we worked on getting privileges and credentials for the hospital, plus regular traveling details like updating passports, etc. Then we went to work on supplies, asking the folks in Ukraine what was needed. The answer? Everything. We had to bring all instruments and plenty of other supplies, including the antibiotics and even the dressings. I quickly realized they didn't have much of anything. So for months we accumulated sutures, drapes and all the ancillary supplies we would need in the operating room. They even asked if I could bring a couple of pacemakers, something I knew nothing about but was able to procure. Eventually, our checklist was complete: drapes, sutures, sponges, pacemakers, thousands of dollars worth of loaned surgical instruments, passports, and, last but not least, a full team.

We flew first to Germany, and as we boarded the plane bound for the Ukraine, I noticed something unusual. There wasn't a long line of people waiting to get on the plane. Practically no one else was getting on this flight. But it was way too late to turn back now, so I nervously boarded.

As we approached the runway, I looked down and saw something I didn't expect: dogs. They were playing on the runway and no one seemed to care. Another indicator that we weren't in Kansas anymore was the back-room negotiating our hosts had to do so we could get through with all of our equipment. I later learned that if those negotiations hadn't been made, the equipment would have been confiscated and sold to the highest bidder. Eventually customs let us through with all of our goods. Praise the Lord! I didn't want to have to tell the hospital that the thousands of dollars of equipment they had loaned to the team was gone.

Odessa was a beautiful city. I could see that at one time it was probably a bustling seaport, but years behind the iron curtain had taken its toll. The people of Odessa seemed to have no expressions on their faces, as if they had no emotion left. The restaurants were mostly empty with the exception of foreigners because no one could afford to go out. I was told the tax rate was over 100 percent. Apparently the rate forced pretty much everyone to cheat on their income tax reporting. It was the way of life.

Our quarters in the Ukraine were an old college dorm. It was plain but comfortable enough. Richard and I shared a room with two fairly small beds. I'm over six feet tall and Richard has several inches on me, so we got used to our feet hanging over the ends of our beds. Not a big deal, all things considered. What did take a little getting used to were the cold showers. This wasn't just room-temperature water. It was Lake Michigan cold, the type that sucks the breath out of your lungs. Not to mention the water that only trickled out of the shower head. Richard and I could always tell when the other was in the shower by the involuntary, high-pitched screams.

Our second day there the host physician came to pick us up and take us to the hospital. While we were waiting for the rest of the team, he asked if I wanted to play chess. There were chess boards everywhere. Now, I would not consider myself a great chess player, but I can hold my own. I was, after all, the secretary of the chess team at a nerdy, all-boys high school. So I thought I'd take him on. Twenty moves later I was so far behind and so destroyed that I gave up. That's when I remembered that Boris Spassky was Russian. I came to the conclusion that all Russians and their offspring must be great at this game. I never played again while I was there. I couldn't stand the humiliation. The rest of the team showed up in time to see my drubbing, and then it was off to the hospital. Time to report to work.

We began to line up the work we would undertake during our stay in the Ukraine. The hospital set us up in a clinic to see the surgical candidates. Language was a barrier, but fortunately some English-speaking students were assigned to us, and that helped. I'm usually able to pick up languages fairly easily. I already spoke French, having lived there in France, and I can get by reasonably well with Spanish. But Romantic languages are much easier to speak than Russian or Ukrainian. There was no hope for me to gain any mastery over these languages in our short time there. So we all stuck to our translators like glue.

The first clinic visit was eye opening. I normally see patients in a private room at my private office, but here they were simply brought into a big, open ward. The patients sat along the wall waiting for their turn. There was no privacy for taking information or for the exam. Also, in order to be seen, the patient had to bring examination gloves for the practitioner. The gloves were not provided by the caregivers. Once the history was taken through a translator, the patient would change clothes and get on the exam table. And yes, the exam was done right there in the middle of the room for everyone to see. I was shocked. Not only was that against the ethical practice in the U.S., I would have lost my entire practice overnight if word got out that exams would be out in the open. But everyone seemed to be OK with it since, I guess, it was a standard practice. In fact, no one even blinked. They were just glad we were there to fix their problems.

My host physician would periodically disappear to do other things. I'm not sure what he did, but I have a strong suspicion.

Remember, the tax rate was over 100 percent, so for him to adequately provide for his family, he probably had to do under-the-radar office procedures to help pay the bills. The wait for standard care was so long that if you wanted it done right away, you paid under the table and expedited your care. The system drove people to cheat. Needless to say, I didn't follow him on these tasks.

After examining and lining up the patients we would care for, it was time to get ready for the operating room. We got the bare minimum of preoperative blood work, and we didn't worry too much about informed consent. Patients simply knew they were not well and needed help, and they were willing for you to do it. The antibiotics that we give before surgery were handed out by us from a bottle that we had brought from the United States. We carried it in our pockets and gave it to patients just before the surgery.

I spoke with the Ukrainian doctor I was paired with through a translator, and he couldn't have been a nicer guy. He had actually been to the States to work with some general gynecologists, but not with an oncologist. His interest was particularly in gynecologic oncology, but he said that in his country the profession did not really exist. There were those in the Ukraine who spent most of their time in that endeavor, but they had no formal training. For that reason, he asked lots of questions about how and why we do things the way we do in the U.S.

Finally, it was time for surgery. When we went to the operating suite, they had us change clothes and go down to the room where there was a team of nurses ready for us to start the cases. They helped us get our gowns on and readied the patients for surgery. My circulating nurse who came with us helped to identify all of the surgical instruments because the person handing us the instruments didn't speak a lick of English.

When we got ready to start the operation, I chose the type of incision that best fit the case. But before I could make the incision, I was told that I couldn't make my standard incision.

"What do you mean I can't do that?" I asked.

"That type of incision is not allowed," the co-surgeon said.

I was shocked. See, when a surgeon starts an operation, the first decisions are a.) how to position the patient, and b.) how to make an incision to best rectify the predicament. For surgery in the pelvic area for a mass or a cancer, the incision is commonly an up-and-down incision. We do that because, if the surgeon needs more room to get the tumor or abnormal growth out, it's an easy conversion to lengthen the incision. You would just take the incision up around the belly button, lengthen it, and voila, you have more room. If you make a bikini incision in the case of a mass, and you end up needing more room, you're officially hosed. You can't get any more room out of that incision. The only other option is to make a second incision.

So I wanted enough room. I did not want to find myself cut short on space and have to compromise the way I would take care of it, which is why I asked my co-surgeon why I couldn't make the incision I wanted. While there was probably something lost in the translation, I eventually came to believe the incision was mandated by hospital policy and not open to discussion. There would be potential fallout from a "wrong" incision. Fallout in a Russian country is not what I wanted. My job was to cut this patient open and try to fix the problem. So I did. Unfortunately, my incision was way too small. As if this wasn't enough, I quickly learned that this particular operating table was broken, and it probably had been that way for eons. Normally a table can be moved up or down and even placed in peculiar positions to accommodate what the surgeon needs. This table had only one position: really low. I'm a fairly tall person, so a table below the waist is a long way down. In surgery, you're cutting and removing things, and it's obviously better to perform those tasks within a certain visual distance. But hey, you adjust. I asked if they could dig a hole in the ground, but they didn't think that was funny. So I shut up about it.

Lighting was another issue. I'm used to a $6,000 light on my head that allows me to see into every nook and cranny in the human abdomen. Because of the nature and locations of the cancers I deal with, I need a light that shines brightly every place my head turns. In this room we had the equivalent of a light bulb, and every time I looked in a nook or cranny, my head got between the light and the object. It would disappear from my sight. That's a problem.

Then there was the suture, of which there was very little. Most of the suture we had was the stash I had brought. So I got about 10 ties out of each suture, even though one was my norm. It wasn't easy, but it was doable. We also were short on sponges, which we use to soak up any blood or hold pressure on areas that bleed a little bit. So to conserve, the staff would simply wring out the blood and use them over and over again. All of this was challenging, but it was still possible to adjust. When you operate in an ivory tower and everything is done for you, you begin to think you can't adjust. But you can.

With each new case, life got easier. We learned how to read hand signals. We learned to appreciate the effort that was put forth by these fine people. Communication was still tricky, but a curse word, of all things, helped make it easier. One day I heard one of my partners say a not-so-nice word in French. I spoke back to him in French to see if he spoke the language, and he did—beautifully. He had learned it while working in Vietnam for a number of years. So the language on the table was French. Here we were, a Russian guy and an American guy operating in the Ukraine and communicating in French. It's a memory that has stuck with me for years.

After all the surgery came the recommendations of what to do next. That part broke my heart. In many of the cancers I deal with, after surgery it's important to deliver radiation or chemotherapy to cure the patients and save lives. But here in the Ukraine, those ancillaries were nearly impossible to get. The patients didn't have the money for the chemotherapy, and the state was not going to provide it. If patients happened to have some money, they could buy the drug on the side and pay the doctor to run it in. Radiation, though, was even more out of the question because the expensive machines that deliver radiation, called linear accelerators, were non-existent. It was a tragedy. I don't know what happened to all the people that we took care of over there, as far as their outcomes and if they were ever able to get any other treatment. From an outsider's point of view, it appeared that if the cancer couldn't be cured surgically, it was hard to hope for any cure at all.

Negotiating our way out of the country started just like the negotiations to get in. The officials didn't want us to leave with our supplies. If they could keep them in customs and extract a ransom, they would do it. I'm sure that money passed under the table to help us get out with the items we brought. Our hosts wanted us to come back, and they wanted us to pass as freely as possible. So eventually we got through customs and onto the plane. We took off late at night, and as we headed into the skies and back to the west, I had plenty of time to reflect on my time in the Ukraine.

I came to realize a few things. Foremost in my mind was how spoiled I am. Things like moving tables and lights seem so small, and yet they're so very big. I take for granted support items like antibiotics and suture. American patients take a lot for granted, as well: not having to bring your own supplies to office visits, the ability to get care in a timely fashion, and, of course, the ability to have privacy with your physician and the health-care team. Those poor ladies lost so much of their dignity.

I also realized that the caregivers I worked with did an incredible job given their constraints. Many of them had been granted the right to train in the Unites States, and yet they still went back to the Ukraine to deliver care. They operated with poor light, broken equipment and less than desirable facilities, but they never complained. My initial lack of sensitivity to their circumstances was, in retrospect, rude. I have a newfound appreciation for what is done in these countries when it comes to their health care. They worked hard, with long hours. They made little pay and were exorbitantly taxed. I tip my hat to all the health-care professionals working in Third World regions.

My time in the Ukraine was a sweet time, a stretching time. It took me out of my comfort zone and taught me to be thankful for all I have. The physicians over there are heroes, and I am the spoiled brat.

Chapter 15

The Unexpected in the OR

Every morning the alarm goes off at the same time. I walk downstairs, make a cup of tea and turn on the coffee maker for my wife. I eat breakfast and read while she checks the e-mails. I shower, make one more cup of tea and then walk out the door. After leaving the garage, I assume the day will be like any other day. It could be a day in the office or maybe in the OR. But sometimes life throws us some surprises. This is a recounting of some of the biggest surprises I've faced in the operating room.

Surprises in the OR are not like surprises in other areas of life. They generally result in lots of heartburn. The operating room is a place to be honored and feared. It's a place where humans put their trust in other humans to hopefully fix that which is broken. There are moments of great satisfaction and moments of great angst. You can be the hero, or you can be the villain. You never know what the outcome will ultimately be, and you certainly don't know how the family will perceive that outcome. Their expectations may not be in line with mine. I've had cases that brought tears to my eyes because I could see the pathology unfold before me, and it wasn't good. I've had cases in which I couldn't wait to talk to the family because the findings were much better than what I had previously thought. Other cases I don't want to talk to the family because I just don't know how to put it into words.

For all those reasons and more, it's a place I treat with great honor. It is a privilege to be there. In fact, one of the few observers I've ever allowed in the OR is my mother, and even then I gave her a set of rules, the most important of which was: Even though she was my mother, I would call the shots. And I also wouldn't talk much so I could stay completely focused. Oh, and she couldn't comment on my singing.

Some cases change in a heartbeat, literally. One such case involving a woman with endometrial cancer still sticks with me. Generally speaking, endometrial cancer cases are pretty straightforward. The patients are often a little on the plump side because obesity is the major cause of endometrial cancer. On paper, the surgery is not as difficult as some of the other cancers we care for. But on this day, we were at the end of the surgery schedule, having previously completed three cancer surgeries. They had all gone well, and, believe it or not, we were still reasonably fresh. I was still in my thirties and very resilient, even with long hours.

In an endometrial cancer case, we remove the uterus, tubes and ovaries, along with some other biopsies of lymph nodes to verify the cancer has not spread. It's an operation that can likely be done with very little blood loss and in a fairly short period of time. This patient was a younger woman with few additional health problems.

The incision was made, and we proceeded with the hysterectomy, which was done in about 45 minutes. At the conclusion of the hysterectomy, I began to remove her lymph nodes. Lymph nodes are little, fatty-looking pieces of tissue that generally sit upon major blood vessels. In endometrial cancer cases, the lymph nodes we take are located down in the pelvis on the blood vessels that feed the legs and the rest of the pelvic structures. These major blood vessels and their accompanying nodes are dissected such that, at the end of the dissection, the vessels sit somewhat bare, and thus are quite visible.

I was cruising along. Blood loss was minimal. The length of time was right on schedule. Then I got a most unexpected request. The anesthesiologist leaned over the table and asked me if I was near the iliac vessels.

"As a matter of fact, I'm right on them," I said.

"Would you feel them to see if you palpate a pulse?" she asked. In other words, she wanted to know if I thought the patient's heart was beating. My eyes immediately flashed to her monitor. All I saw was a flat line. Since I was right on the vessels, feeling them wasn't a problem. So I did. No heartbeat. Her heart had come to a dead standstill, no pun intended. She was lying there, in a sense, dying.

I turned to my resident. "Jump on the table and start chest compressions," I said. He stared at me, stunned, so I told him I wasn't kidding. Then he complied. I turned to the anesthesiologist and asked for backup in the room. I covered up the wound a bit and watched my resident give the compressions. My hand was on her iliac, and I could tell that blood was perfusing through her body with the chest compressions. In seconds the room was filled with anesthesiologists. The cavalry had arrived. But, to be honest, even with all of those MDs in the room, the most important person was the resident performing the chest compressions. He was keeping her alive. The anesthesiologists, like myself, were all baffled. None of us had a clue as to why her heart stopped. For a few seconds she started to turn a little blue.

I was beyond nervous. Here I was operating on a young woman with a curable cancer. Everything had gone by the book. She was trusting me to provide her the best care I could, and, as far as I knew, I was doing that. And yet, the dam broke. I was sweating bullets. I began to pray. I needed divine intervention.

Thirty seconds went by, then 60, then two minutes. My resident was doing the best he could. Anesthesia was doing everything to reverse every drug they had given her. I didn't know what would happen. Then, about three minutes into this awful experience, her heart started to beat. At first it was slow, and then it sped up to a normal rate. I still had my hand on her iliac, and this time without my resident pushing on her I could feel the pulse of blood heading down to her legs. I didn't know whether to celebrate or hold my breath. I just wanted the beat to go on. One minute, two minutes--the time ticked on and so did the heart. Her bluish color began to turn a little more pink. I wasn't in the mood to operate anymore. I just wanted to feel the blood rush down to her legs and thank the Lord. She was alive. Was she all there mentally? Hopefully. I decided to expedite the case, finish up, and get out of Dodge. It was one of my faster closings of the abdomen. It's amazing how you can fly when you have to.

She woke up at the end of the surgery and was mentally perfect. There were absolutely no residual effects of what had happened. I told her family, and we were all baffled. Even the anesthesiologists had no clue. It was just one of those weird moments. It has never happened to me since.

Another surprising case involved a patient who had cervical cancer. When women develop that cancer, they either undergo radiation and chemotherapy or they have a surgery called a radical hysterectomy. A radical hysterectomy takes about two and a half hours, which is about an hour longer than a standard hysterectomy.

Many cervical cancer patients are young women. My mentor and former professor taught his students to make very cosmetically pleasing closures on these patients. Actually, he insisted on this with every patient, regardless of age. As he would say, "When the patient looks in the mirror years later, the only thing she remembers is that incision. Make it pretty." We were never allowed to take any shortcuts on the closure of the skin. It had to be done as the plastic surgery folks would do. No exceptions. Placing staples is a whole lot faster, but, in the end, it truly doesn't look as nice. That was important to him, and I honor that *almost* always. This day was the one exception.

I was about an hour and thirty minutes into the surgery when someone walked into the OR and said there was a bad storm brewing outside. The OR was in the heart of a large building, so you couldn't hear the rain fall, and there were no windows. No more than 10 minutes later, a lightning bolt must have hit a large transformer and knocked out all of the hospital's power. Now, don't think that hospitals haven't thought of this. They all have backup generators. Once the power goes down, generators kick on within about a second. There are so many people on breathing machines or other support devices that an electrical outing would be catastrophic.

So the lights went out in our room, as they did in every room, but as all the lights came back on, the fuse box for our room must have blown. The lights went out a second time. Lightning, in a sense, struck twice. I was in a virtual blackout. The anesthesiologist immediately started breathing for the patient manually with a type of bag. He also was able to monitor the patient manually as far as her vital signs. But the kicker was that I had a patient open, and I was way past the point of no return. I couldn't just quit and resume later. The uterus had lost all of its blood supply and was not looking too healthy, yet it wasn't out. I needed about 10 more minutes. And then I would need to place some drains and close the abdomen. The other problem was that there was not another obvious OR room that I could move into. They were all occupied or were not clean or set up properly. So the head of the OR came in to help me figure out our options. I gave him the approximate time for completing the procedure. With that, we decided the most expeditious route would be to bring in some flashlights and shine them into the belly. They quickly identified about five flashlights and brought them in with people to hold them. I was then asked kindly to hurry up. At that moment, I felt like I was operating in some Third World country. In fact, I've actually experienced that before, and this was not far off.

So we made haste, completed the procedure and then tackled the closure. When I got to the skin, I simply assumed that I would, on this one day, not make a plastics closure. My boss knew of the unusual events that had occurred. When I spoke with him on the phone, I didn't even tell him that I might amend the way I normally close the skin. I thought that was just assumed. After all, I was operating without electricity. So staples it was, and we were out of there.

The patient did fine. They fixed the fuse box. Life was back to normal. The next day, my pager went off. I was summoned to the office of my mentor and professor. He asked why I had used staples on the patient. I reminded him that this was the patient on whom I had operated in the dark. Plus, the head of the OR had told me to make it snappy. But my professor merely reminded me that this was not our standard closure.

What do I say to that? It wasn't the way, and I had altered the way. I hated to disappoint him because I respected him so much. Still, I *did* feel like the circumstances had changed. But I just said it would never happen again. Even if the lights went out. But I knew that wouldn't happen. How often does lightning strike twice in the same OR?

I approach my final anecdote of the unexpected with great trepidation. If I had to look back and recount the toughest five days of my life, this day would be up there. It brought many tears to my eyes, which, for me, is a rarity. It's a story I didn't really want to tell, but if I am describing a life of medicine and all the joy it brings, it's only fair to mention the great sadness that can accompany it, as well.

A patient--"Mrs. Smith"--was sent to me for a pelvic mass. Her CT scan showed it likely to be an ovarian cancer, and her exam suggested the same. Mrs. Smith and her family were some of the nicest people I had ever encountered in my field. She was around 70 years old and had a few additional medical problems. A heart problem forced her to take blood thinners, which make it easier for a patient to bleed during surgery. So before we could operate, the blood thinner had to be reversed. It's a tricky business. (My father actually died from a complication of this issue.) The family was fine with waiting an extra week for the surgery so the reversal process could be completed as needed. Her husband seemed to be a wonderful man, and it was apparent the couple was quite happily married. They understood what faced them and were ready to take it on.

The day arrived and I took Mrs. Smith back to the operating room. Having been there thousands of times before, I assumed this day would be like any other. The approach on a case like this is fairly standard. Of course, there are nuances to every patient, but all in all, they're usually similar. I made an incision into the abdomen and noted the large mass in the ovary. The mass was removed in short order, and I had it sent out for what we call a "frozen section," in which the pathologist opens up the ovary and literally freezes a portion to examine under a microscope. The diagnosis was quite easy, as this was, indeed, an ovarian cancer. Once the diagnosis is made, one of two things can happen. The practitioner looks around, and if there is no other sign of cancer, a number of biopsies are performed around the abdomen to determine whether the cancer has spread. If, on the other hand, there are signs of spread, then the goal of the procedure becomes an attempt to remove all of the cancer. Sometimes patients are in between these two points--you see some spread of cancer, but not a lot. That is where Mrs. Smith was on that day. When that scenario presents itself, then the goal is to undertake both of the procedures described (i.e. get the cancer out, and also see how far it has spread). So that was my mission.

One part of this operation is the removal of lymph nodes. As I mentioned previously, lymph nodes normally sit right on top of major blood vessels, and when you operate on top of blood vessels you can potentially injure them. The larger blood vessels will create an increased chance for the patient to have significant bleeding. Veins are also much worse than arteries because they are so hard to fix.

There's another blood-related hurdle of cancer surgery. You see, in order to grow, a cancer needs its own source of blood. Since blood vessels don't usually exist for a cancer, the cancer figures out how to get them. It produces proteins that tell the human body to grow new blood vessels and send them to the cancer. We call this process neovascularization. This process can be unnerving as a surgeon because these new blood vessels are not listed in the textbook. Even if you stayed awake in class, this blood supply was never mentioned since it's not supposed to be there. So, if it's not supposed to be there, how is a surgeon supposed to find it? Well, part of the answer is that we learn to anticipate it. But even if we try to anticipate the blood vessels, they can hide. Sometimes you can't see them until you're in them. That is a major risk to cancer surgery.

Anyway, Mrs. Smith's case was proceeding well and my team and I were closing in on the last part, which is to take the lymph nodes off of the vena cava (the largest vein in the human body). In any cancer operation that requires the removal of lymph nodes, each particular cancer has, in essence, its own lymph nodes, which in this case, were located directly under the belly button. This is not an uncommon location for an ovarian cancer to spread from time to time.

While staring into that region, I began to lift up this particular node, grasping it with a type of clamp that is designed to elevate it but not crush it. When I picked up the node, it seemed a little stuck. Normally the node would lift up with relative ease. But on this day it seemed a bit glued to the vena cava. I began to manipulate it when horror of horrors struck. Blood began gushing out at an alarming rate. There is no worse sight in an operating arena than blood pouring from a source that cannot be easily stopped. It was scarily obvious that the vena cava had a sizable hole. I looked at Brenda, my longtime assistant, and she read me like a book. Brenda and I had performed more than 3,000 cases together. She seemed to know my thoughts even before I did. She was, and is, a true professional. Brenda saw the tension in my face and knew something was very wrong.

I immediately put pressure on the bleeder with an instrument called a sponge stick, which is kind of like a finger in the dike. It can buy you valuable time--something I desperately needed at that moment. The bleeding was temporarily controlled. I then got everything organized and made my visibility of the wound as best as it could be. Remember, I was working in a hole in the belly, and at the very bottom of that hole was an even smaller hole in this blood vessel. In order to see the smaller hole, the operating area has to be big enough to make sure that little hole is perfectly visible. It also helps to have a well-lit environment, so my headlight, which is very powerful, and the overhead lights were all adjusted to make sure there was enough light to fix the problem. Next, we moved other things, such as intestines, well out of the way. Then it was time to call for help. This hole in the vena cava still had a lymph node stuck to it that didn't want to let go. If I pulled on it anymore it would just make the hole that much bigger and that much harder to fix. I was between a rock and a hard place, and I knew that outside assistance could only help the situation. My next-door neighbor in the OR is the vascular team, and I have always loved having them close by, as this contingency is an ever present possibility. I know that, and they know that. So I called in the troops.

The initial blood loss was probably about 200 to 300 ccs. In the big picture, that's not a lot. I had anesthesia informed of the situation and held pressure until the vascular group arrived. I was hopeful this could be resolved, but I wanted to be anywhere but there at that moment.

As the vascular surgeon and I proceeded, it became apparent that the defect in the vessel was significant. We tried holding pressure upstream and downstream to gain control of the bleeding until the vessel could be sewn. Unfortunately, she had vessels coming up from underneath the cava that caused a persistent flow of blood. With blood constantly in the field of surgery, sewing this vein was exceedingly cumbersome. It would be like trying to fix a backed-up sink, and every time you take your hand away from the drain to fix it, the sink immediately fills up with water. But unlike water, you couldn't see through the red of the blood. The other hurdle was that we needed to get all other tissue (i.e. lymph nodes) cleared off the vein to see it clearly. In the midst of all of this, the patient lost more blood, which was quickly replaced, but when enough blood is lost, patients begin to lose their ability to clot off any bleeding. This allows for bleeding from other sources. So now she was not only bleeding from this vessel, but also from other locations where I had taken the cancer out. We were replacing other blood products in hopes of stopping the extraneous bleeding, but it was difficult to stem the tide.

After several hours of work, we finally gained control of the original defect in the big vein. But I knew that in an elderly patient with a number of medical problems, not to mention an advanced cancer, the deck was stacked against us. I hadn't had time to dwell on the ancillary issues up until this point because our focus was so intense on the problem at hand. But with this bleeding stopped, I began to turn my attention to other issues, such as addressing the family and figuring out where to go from here. She was alive, but not strong. I believed in my heart that she would not likely pull through. It was my problem and my burden to bear. I had to deal with it.

We closed her up and sent her to the ICU. I feared that at any time they would call me to say she had started bleeding again. I didn't know if I was on borrowed time. I left the room and went to work on the rest of the details.

When I got out of the OR, I had an overwhelmingly heavy cross to bear. This woman and her family had entrusted her life to me, and I felt like I had failed. I fully realized that it was a difficult case and that I am only human, but that doesn't make it easier. I tried to gather up the strength to talk with the family.

As I entered the room, the place was filled with those who loved her. I took a deep breath, said a silent prayer and hoped that I could contain my emotion long enough to explain where things stood. I tried as hard as I knew how, but the tears began to well up. I am sure that my voice cracked, but I tried to remain strong for the family. I explained everything as best as I could. I wanted them to understand the precarious situation. I wanted them to know how tenuous life was at present. They knew. They understood. I was fortunate because they had a grasp on how life does not depend on the hands of man but on the hands of God. After I answered all of the questions, I asked if I could pray with them. We did. And they knew that this life was presently in the balance. They asked me to do everything that I could, and I gave them my word that I would. I stayed at Mrs. Smith's bedside and manipulated every variable that I could to give her the best chance of survival.

Three hours later she died. I again talked with the family and gave them my condolences. I didn't know what else to say. I simply wanted to crawl into a hole and die. My worst nightmare had come true. I had always known this was a possibility, but no one can prepare you for the devastating consequences. It shakes you to the very core. It's a scar that is not easily repaired.

I went to my office. It was late at night, and I was there all alone. I broke down and sobbed. I utterly wept. It is next to impossible to not take it personally. A life was entrusted to me, and that life was no more. It was a profound sense of failure. My job is to preserve, prolong, or perhaps save a life, and that did not happen.

I couldn't wait to get home and be held by my wife. Better yet, I wanted to put it on her shoulders. When I'm alone in that state, it's even more painful. I didn't even need her to *fully* understand the pain and suffering; I just needed her to listen and hold me.

Looking back on that day still affects me. Some people have bad days that consist of a quarrel or a headache. And while I'm not belittling those things, I realized that day that if I had an unexpectedly bad day, someone could die. That's a hard truth for me to wrap my head around. But I signed up for the job. And I am not God. I am just a man. I do the best I can, for that is all any of us can do.

Chapter 16

The Pursuit of Knowledge

I have been teaching students and residents for more than 20 years. The longer I do it, the more I enjoy it. Part of the reason for that satisfaction is that, as I pour a small piece of myself into these future physicians, I then see them go out into the community and care for others. Watching them develop as professionals, and as people, makes me proud.

If there is anything redeeming about what I do for a living or what I have done, I owe many thanks to those who prepared me for it. There are countless people I could thank for sacrificing themselves to help me achieve my dreams, but, to narrow it down, four professors in particular inspired me in different ways. The principles they taught were rather simple but nonetheless profound. Through them, I learned the importance of pursuing knowledge, as well as the qualities of stretching yourself professionally, tenacity and consistency.

Intelligence is innate to a certain extent, and yet it can be partially earned. Even Einstein was not bright when he was born. He had a mind that allowed him to learn at a faster rate and grasp more than the rest of us, but he still had to apply himself. I became enamored with raw intelligence when I was exposed to it during my training. I realized that without knowledge, I could not save lives. Without learning and retaining information, patients could lose their lives because I wouldn't be able to figure out the puzzles in front of me. These patients needed me to figure it out.

Learning in college was just to make grades so I could get into medical school. I loved college and all its trappings, but

I didn't honestly think it was life changing. I wasn't fixing things or helping people. I wasn't inventing things. In my first two years of medical school, though, it seemed like these classes could have some impact. Still, I didn't fully realize the importance of pursuing knowledge until the third year of medical school. That's the year where you actually walk into rooms with real people and ask them real questions to (hopefully) really help them. It was a wake-up call. If I lacked knowledge, bad things could happen. Even *with* knowledge, bad things could happen, but at least I could better predict it. It's like an engineer who designs bridges. If he doesn't know how much weight the bridge can hold, people can lose their lives. So pursuing knowledge is important for all of us, regardless of our calling. I was just a late bloomer when it came to figuring that out.

At Southwestern Medical School in Dallas, the Internal Medicine Department had a way of drilling that point home. My father had trained in that department 30 years before I arrived on the scene, and the chairman of the department had more of an impact on me than any other professor. I'm sure he wouldn't remember me. I was just one of 200 students who came through the service every year. But even as a busy chairman, he still made time to teach.

The chairman was a nephrologist by training, meaning he was a kidney specialist. He had a commanding presence—someone who seemed to know all the answers. I thought he was brilliant, and he intimidated the tar out of me. During his presentations, he would walk around the room. Or, more accurately, he would prowl. He was like a lion stalking his prey. In this case, the prey were those who didn't know the answer. As he paced, he would stop behind various desks and ask general questions. As we got deeper into the case, the questions became more and more difficult.

You never knew where he was going to stop as he paced. It was like musical chairs. Round and round he would go, walking behind the desks. He would slow down behind several desks until he had chosen just the right one. He had a sixth sense about who knew the answer and who didn't, like St. Nick knowing who was naughty and who was nice. I'm convinced of it. If you knew the answer, he could tell from the swagger in your posture.

One day in my third year, during a lecture to a small group of students from my class, he stopped at my desk and placed his hands on my shoulders.

"So," he said as he walked around to the front of my desk to look at my name tag, "Dr. Puls, Mrs. Smith's potassium is low. We have to fix it. But we need to know why it is low. So...why is it so low?"

At that moment it felt like the whole world's eyes were on me. I thought I was going to die or pee my pants. I didn't have the slightest clue why Mrs. Smith's potassium was low. He knew that, and I knew that. My world was falling apart. All of my friends would know that I was stupid. They would all know that I was not able to save this poor lady's life. There was deafening silence for about 30 seconds.

"Dr. Puls, do you know?" he repeated.

I needed to come clean. It was time to put myself—and everyone else for that matter—out of agony. In the most quiet and humble voice I could muster, I simply said, "I don't know."

He walked around the desk to stare me down again. I don't know how long he stared, but it felt like an eternity. I wanted to crawl under my desk and melt away.

"So, Dr. Puls, you don't know," he said, then reached into his pocket. I had no idea where this was going. Part of me hoped he would just shoot me to put me out of my misery. Instead, he pulled out a quarter and laid it on my desk.

"Here's a quarter," he said. "Call your grandmother, I'm sure she'll know the answer."

There was a lot of laughter—all at my expense, of course. I wanted to get up and run out of the room, but it was time to suck it up and take it. The laughter only lasted a couple seconds, though, because once he looked up and stared around the room, the lion was on the prowl again. Trust me, there were other people in that room who didn't know the answer.

But you know, in the end, I learned the answer. And, perhaps more importantly, I learned *why* I needed to know the answer. My professor's intimidation tactics were not really serious, but they taught me that raw knowledge and intelligence was worthy of pursuit. After that year I began to change my approach to medicine. I understood that without certain knowledge, I could put lives in jeopardy, and that was worth memorizing millions of facts. To this day I have a stack of about 500 cards in my desk filled with tons of facts I spent years memorizing.

After graduating, they let me matriculate from medical school into residency, and the transition was radical. I went from studying all the time to working and studying all the time. When I went through training, the 80-hour work week cap was not in place. Thus, many weeks I'd work more than a hundred hours. The reason that could happen was that, during nights on call, a resident would work the day before and the day after. This would be, in essence, a 36-hour shift. After a while we got used to it, but I can't say I never fell asleep in a patient's room. There were definitely a few times when patients with monotone voices would talk slowly and quietly, and it was just like a lullaby. I'm not sure if a patient ever caught me, but I wouldn't be surprised. Fortunately it was usually too hectic for something like that to occur.

When I started my residency I seemed to have a kinship with one of my professors. She was a maternal-fetal medicine (MFM) specialist, which is someone who specializes in high-risk pregnancies. She and I just hit it off. I loved to teach younger residents, and she loved to teach too. She had a great grasp of knowledge, and I wanted a great grasp of knowledge.

After my first year of residency, I won some teaching awards from the students, and she seemed to be pleased by that. So she called me to her office to talk. At first I thought I had really messed up. A call to the office of the attending physician is usually not a good sign. But this one was good. And it was inspiring. She asked what I was going to do when I finished training, and I said that I wanted to go into private practice in general OB-GYN. That's when she challenged me to think a little differently and explained to me what's involved in doing a fellowship. She also challenged me to consider teaching long term. I didn't know what to say. I thought about it and talked it over with my wife. I had never really considered that kind of path. But she and I discussed it often, and eventually I began to look at all of the options to decide if any of them were right for my family and me. Would I stop at four years training at the end of residency or would I go on? If I went on, what else would I do? There were multiple different fields that I could have done with oncology being only one of them. All of my friends would be getting out into the real world. That was so tempting because for once they could set their own schedule and determine how hard they wanted to work. I wanted to do that too. But this kind of extra commitment to go on with more training would mean in essence starting all over, yet again. "Are you crazy?" I would say to myself. "Just go out and finish and live this life before you," I would say. Then my alter ego would tell me to listen to my professor. What should I do? I must have vacillated on this decision 500 times. This process dragged on for about a year until I finally decided she was right. Now I didn't choose her area of specialty, but I chose the one I have spent my career doing, which is oncology. I profoundly thanked her for encouraging me to consider a path less taken. She stretched my thought process. She pushed to think bigger than what I had thought. I didn't honestly know if I was bright enough or driven enough to do a fellowship. But I immersed myself into the process of trying to get a job. There were very few positions available nationally. But every time I got on a plane to go interview and of course, spends thousands of dollars I didn't have, to try and convince someone who didn't know me to give me job, I thought of her.

    My biggest sadness in all of this was that she never saw her encouragement for me to push myself, come to fruition. This woman loved to travel. She had several friends that traveled with her all over the world. In the fall of my third year of residency she told me the details of her "next" big trip. She and her friends were going Rwanda to see gorillas in their native country. But the small commuter plane on which she was traveling went down in the mountains of Rwanda and she didn't survive. It was a tragic end to a dear life. She pushed and stretched me and taught me to believe in myself professionally. I could never have thanked her enough.

As I said, I spent what little money there was from residency (I made about 3 dollars an hour considering the 100 hour work week), flying all over kingdom come to get a job. Ultimately it paid off after two years of trying. That life-changing day came; I was offered one of the few jobs in the country in gynecologic oncology. I'll remember that day forever. It was like a dream. I was actually going to be a gynecologic oncologist. I realize most people probably look at what I do for a living and think I've lost my mind. But I was wired differently. And that day was truly one of the most life changing days of my life. I can still remember the phone call and the offer. I took the job without hesitation. I was pushed to stretch myself by one of mentors, and barring something unforeseeable, it was going to happen. I was again in debt to someone who poured their life into me.

So I moved to Kentucky and started a new job-a fellowship in gynecologic oncology. Once again, I was at the bottom of the totem pole—a position I was getting quite used to. With this new job came my next major life lesson. (I wasn't anticipating anymore new lessons.) When I landed at the University of Kentucky I met Paul. Paul was a fellow in oncology like myself except that he was a year ahead of me. But with this new experience in training, a new lesson was upon me. Paul not only knew more than I'll ever know, he always out-worked me, as well. Paul grew up about as poor as you can imagine, but early in life he came to the conclusion that if you persevere, you can succeed.

Paul was at the top of every class he was ever a part of. Even kindergarten, I'm sure. He was smarter than any contemporary I've ever had. He always knew information before I did, which earned him the nickname Radar. One time I was standing at a patient's bedside while she was dying, and I called to let him know. He promptly replied that he already knew. I asked how that was even possible, since I was at her bedside and he wasn't. Well, he was talking to the patient's nurse at that exact moment on another phone. I looked down toward the nurse, and sure enough, it was true.

Here's just a brief rundown of some of Paul's engagements and accomplishments: He taught surgery and medicine to residents and fellows; he ran a 35-acre working farm, in which he not only cared for his immediate family, but also his parents and sometimes his sister (and, of course, the animals); he ran the entire operating room facilities (30-plus ORs); he was on the editorial board of our national journal; he published papers; he ran research projects; and he was heavily involved at his church. Oh, and then he went back to Harvard for a graduate degree. Although he won't say, he probably made all As. He was a true renaissance man.

To top it off, Paul was humble beyond belief. He was so perfect it could drive you insane. I realized the first week I worked with him that I could never equal him, so I gave up trying. But the most powerful, valuable thing he taught me was tenacity. From him I learned that if you apply yourself and work harder than everybody else, you can master something. And it wasn't about competition. Paul just wanted to take great care of people, and to do this he knew he had to work hard. So I learned that even from my starting point (i.e. that of someone a lot less brilliant) I could accomplish the tasks before me by employing a solid work ethic. He was a friend and inspiration during that part of my life. Even now he continues to be an inspiration long after we finished working together as co-fellows.

My final lesson came from my mentor and professor during my fellowship years. His fellows called him Captain Jack, and Captain Jack was the man because he's the guy who took a chance on me. He gave me the opportunity to become a gynecologic oncologist, and because of that I dedicated myself to doing the work he laid before me. I took call essentially every night for several years because that is the way he wanted it, and I was happy to accommodate. I wanted to honor the man who was training me.

Captain Jack pushed me on some lessons I mentioned previously: pursuing knowledge and being tenacious. But the biggest lesson he taught me was consistency. Captain Jack developed a method in the operating room that seemed to always work, and with minimum blood loss. We affectionately called it "the way." The way also carried over into our methods of caring for patients on the floor. As Captain Jack would say, "it may not be the only way, but it's a very good way and it will keep you out of trouble".

There are about 50 different ways to do surgery, but he was into the etiquette and tradition of surgery and all the meticulous aspects of it. It may not seem like much to most people—even surgeons, for that matter—but even how you tie your knots on sutures is part of "the way." The first surgical lesson I learned when I started operating was how to tie knots. Yes, I technically learned to tie knots in residency, but it wasn't necessarily pretty. The comparison I give to my residents is that you can teach your kids to chew their food with their mouths open, and it's an OK way of getting the job done, but it's not too efficient, and it certainly isn't pretty to look at. Well, that's like tying knots. If it's done correctly and consistently, it will always keep you out of trouble. Consistency was the name of the game. We held every instrument correctly. We tied every knot perfectly. There were no exceptions. Ever. That was the way.

I reflected on this consistency. This man had developed one of the best national reputations I had ever seen. People came in from all over to have this man care for them. So I decided that if a commitment to consistency had earned him this kind of reputation and consistently good outcomes, then it must be worthy of pursuing. And here I am, many years out of my training, and what do I do? I do it the way. I teach my residents to do it the way, and I don't compromise. I see that it is tried and true. It has been a good way, and it has served a whole new generation of young physicians well. It doesn't allow for cutting corners, and it consistently keeps us out of trouble. It is, after all, "the way". Everything else is the other way.

Those four professors have affected the way I think, the way I pursue knowledge, and the way I see the future of my education and others' education. The lessons these people have taught me have served me well, and I'd argue that they would serve anyone well. All of these mentors owed me nothing, and at the end I owed them everything. I am trying to be diligent in teaching these lessons to the next generation, and yet none of my former professors know that I am writing these comments about them. That's the way it usually goes with teachers. They pour themselves into younger people and never expect anything in return. It is a selfless type of giving. Those who will need healthcare in the next generation will reap the rewards of those lessons, as well. Because of a good teacher, those patients will have better experiences with healthcare professionals.

So I tip my hat to teachers. Without them, none of what we achieve would be possible.

Made in the USA
Lexington, KY
20 August 2010